Taylor's Weekend Gardening Guides

Barbara Ellis, Editor
Frances Tenenbaum, Series Editor

HOUGHTON MIFFLIN COMPANY
Boston • New York 1997

Organic Pest & Disease Control

How to Grow a Healthy, Problem-Free Garden

Copyright © 1997 by Houghton Mifflin Company
Drawings copyright © 1997 by Steve Buchanan

All rights reserved

For information about permission to reproduce selections from this book, write to Permissions, Houghton Mifflin Company, 215 Park Avenue South, New York, New York 10003.

For information about this and other Houghton Mifflin trade and reference books and multimedia products, visit The Bookstore at Houghton Mifflin on the World Wide Web at http://www.hmco.com/trade/.

Taylor's Guide is a registered trademark of Houghton Mifflin Company.

Library of Congress Cataloging-in-Publication Data

Organic pest & disease control: how to grow a healthy, problem-free garden / Barbara Ellis, editor.
      p. cm. — (Taylor's weekend gardening guides)
  Includes bibliographical references (p.   ) and index.
  ISBN 0-395-81370-0 (pbk.)
  1. Garden pests — Biological control. 2. Organic gardening. 3. Plant diseases. 4. Plants, Protection of. I. Ellis, Barbara W. II. Series.
SB974.0725 1997
635.9'296 — dc20   97-14782

Printed in the United States of America

RMT 10 9 8 7 6 5 4 3 2 1

Book design by Deborah Fillion
Cover photograph © by Dwight R. Kuhn

Contents

Introduction 1

Chapter 1 — Creating a Healthy Garden 3
Eight Steps to a Healthy Garden 4
Build the Soil 4
Choose the Right Plants 8
Plan Diverse Plantings 10
Buy Healthy Plants 15
Provide Proper Plant Care 16
Keep Garden Records 20
Scout for Problems 20
Choose the Right Controls 24

Chapter 2 — A Guide to Garden Pests 35
From Aphids to Wireworms 36

Chapter 3 — A Guide to Garden Diseases 69
From Apple Scab to Wilts 70

Chapter 4 — Dealing with Weeds 85
Preventing Weed Problems 85
Controlling Garden Weeds 87
From Bindweeds to Yellow Nutsedge 90

Chapter 5 — Handling Animal Pests 109
Dealing with Deer 109
Keeping Out Rabbits 110
Managing Mice, Voles, and Moles 110
Baffling Birds 111
Handling Other Animal Pests 111

Hardiness Zone Map 115
Sources 116
Credits 117
Index 118

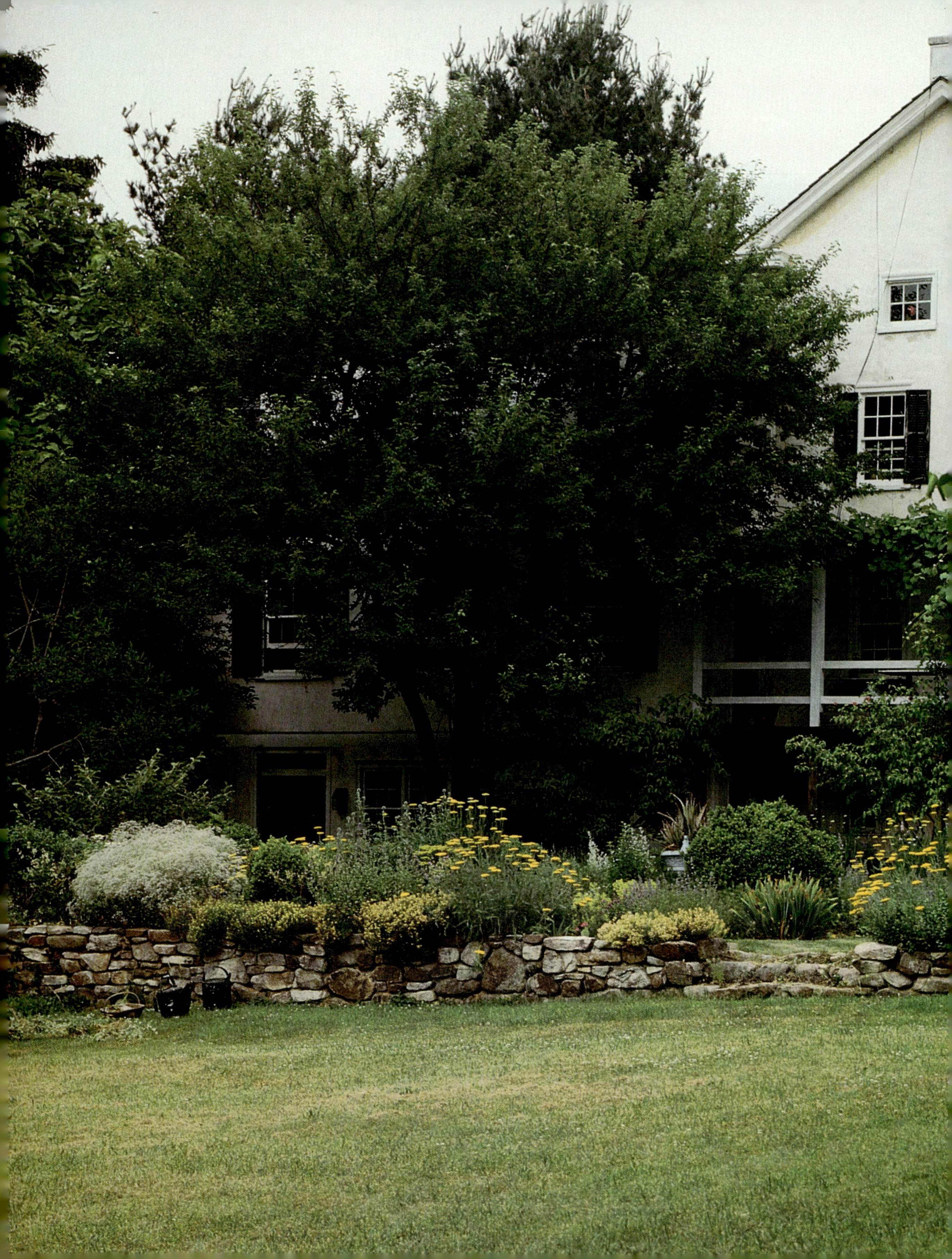

Pests, diseases, and weeds are among the most frustrating problems gardeners encounter. Fortunately, there are simple, effective ways to keep them in check—without resorting to dangerous chemicals. This all-organic book provides practical information on how to grow a garden that stays healthy naturally. You'll learn simple techniques for creating a healthy, thriving garden that has built-in defenses against insects and diseases, as well as ways to encourage a host of allies that are only too happy to help rid your garden of pesky problems. For help with specific problems, turn to the encyclopedia entries on common pests, diseases, weeds, and animals.

Planting a mixture of herbs and flowers is an ideal way to invite beneficial insects to your garden. This raised bed, located in full sun, provides the well-drained soil conditions most herbs prefer.

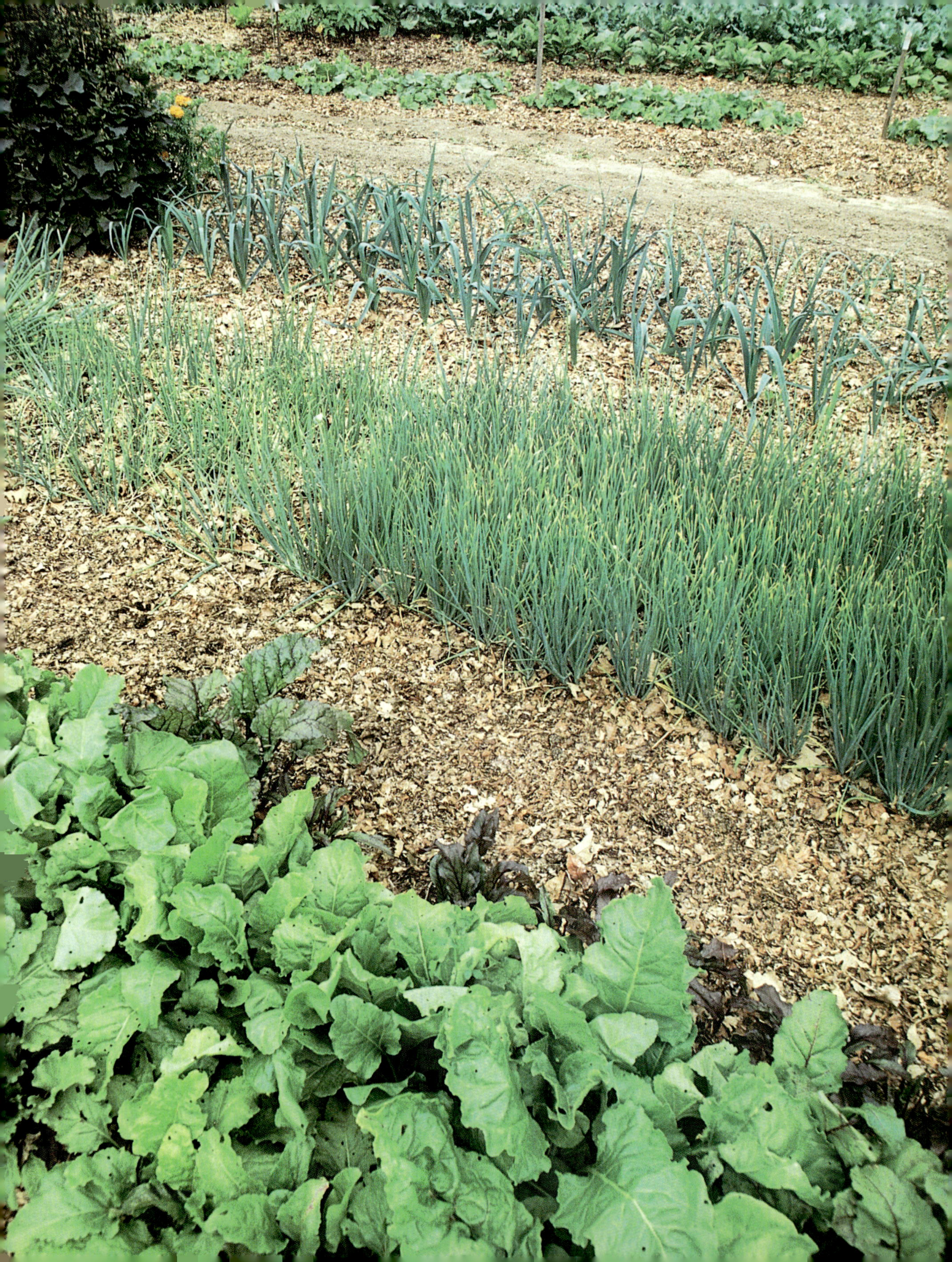

Chapter 1:
Creating a Healthy Garden

Building a healthy garden has much more to do with controlling pests and diseases than you might think. That's because basic good gardening practices — like adding organic matter to soil, spreading mulch, pruning properly, and even putting up birdhouses and planting flowers for beneficial insects to feast on — encourage vigorous, healthy plants. And healthy plants are far less prone to pest and disease problems than weak, struggling ones. In fact, many insects and diseases are programmed by nature to colonize weak plants. Vigorous, healthy plants also are much more able to shrug off problems that do occur.

Good gardening practices are the cornerstones of organic gardening. Instead of waging war against pests and diseases with an arsenal of chemicals, organic gardeners seek a partnership with nature. They attempt to nudge the ecosystem into a healthy balance rather than bludgeon it into submission. Preventive techniques like building healthy soil are important first lines of defense against pests.

Mulching with organic matter, such as chopped leaves, shredded bark, or compost, helps provide optimum growing conditions by adding organic matter to the soil and keeping the soil moist. It also helps prevent competition from weeds.

Beneficial insects and animals are welcome allies. "Soft" controls, such as soaps, oils, and beneficial organisms like BT *(Bacillus thuringiensis),* are used before stronger, plant-derived sprays and dusts, such as rotenone and sabadilla.

EIGHT STEPS TO A HEALTHY GARDEN

This chapter provides an overview of the safe, organic approach to pest and disease control. You can use the eight, easy-to-remember steps to guide you in building a healthy garden full of vigorous, thriving plants. While there aren't any foolproof ways to avoid *all* garden problems, these steps will show you how to give your plants the best possible chances.

1. Build the Soil
2. Choose the Right Plants
3. Plan Diverse Plantings
4. Buy Healthy Plants
5. Provide Proper Plant Care
6. Keep Garden Records
7. Scout for Problems
8. Choose the Right Controls

You are probably already doing many of these things, without even realizing that you're helping to prevent pest and disease problems. But when you take advantage of all the steps, you'll provide the right conditions for a landscape that is naturally healthy, vigorous, productive, and beautiful.

BUILD THE SOIL

To grow healthy plants you need to provide them with healthy soil. Few gardeners are blessed with ideal soil conditions to start with, but it is possible to improve nearly every type of soil with a little work. Building soil that is loose enough to allow roots to spread easily and has ample air space so they can get plenty of oxygen is a primary goal. For vigorous growth, plants also need soil that holds an adequate supply of water and dissolved nutrients. Healthy soil also supports a huge and diverse population of organisms, ranging from microscopic bacteria to wriggling worms. These organisms play an important role in providing good growing conditions: they recycle nutrients, mix and churn soil particles, improve soil structure, and actually help control some pests and diseases.

Knowing the soil conditions you have to start with will tell you what steps you need to take to build better soil. Before beginning any new plantings, spend some time observing and assessing the soil. Soil texture and drainage are quite easy to evaluate with the simple tests below. You can test your soil for acidity and nutrients at home or, for more accurate results, send a sample to a state or private testing lab.

Simple Soil Tests. An easy way to assess your soil's texture (its sand, silt, and clay content) is to take a handful of moist earth and feel how it responds when you rub and squeeze it. If it feels gritty, it is high in sand. If it feels sticky, it's high

Soil Texture Test

You can use a quart jar with a tightly fitting lid to determine your soil's texture — the percentage of sand, silt, and clay it contains. Take a sample of soil, remove rocks and organic matter, and pulverize it. Add water to the jar (about two-thirds full), add the soil sample, and a teaspoon of nonsudsing dishwasher detergent. Shake for 10 to 15 minutes, then watch as the soil particles settle. The sand will settle first, followed by the silt and clay. Measure each layer of particles as it settles, in the time frame indicated. Then divide the depth of each layer by the total depth of the soil in the jar, and multiply by 100. The number you get will be a percentage that indicates how much sand, silt, or clay is in your soil. Loam soil, which has 40 percent each of sand and silt, and 20 percent clay is ideal for gardening. Clay soil contains 60 percent clay and 20 percent each of sand and silt.

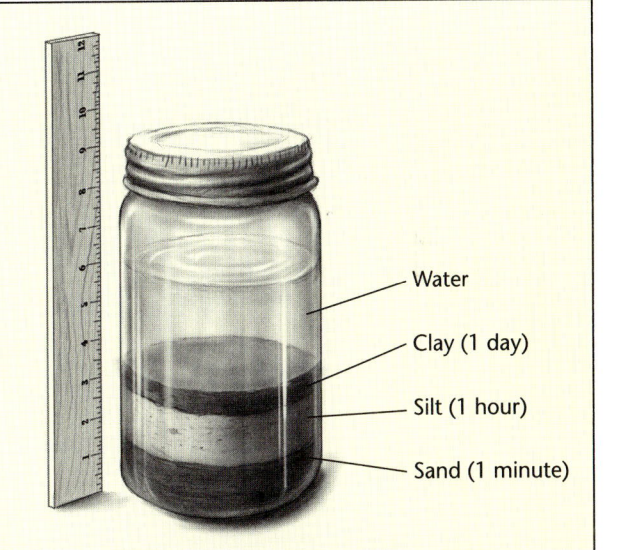

in clay. Sandy soils tend to drain quickly, and they benefit from added organic matter (such as compost) to help retain water and nutrients for healthy plant growth. Clay soils, on the other hand, tend to hold too much water, promoting root rots and other disease problems. Adding organic matter here, too, will help, by loosening the soil and improving drainage.

To test the soil's overall drainage, dig one or more test holes about 2 feet deep. Fill the hole(s) with water and check to see if any water remains in the bottom of the hole after 24 hours. If all the water is gone, the drainage is fine. If just a little water remains, the drainage will be adequate for some plants but not for all. If most of the water remains, drainage is so poor that many plants will suffer.

Whether you test at home or at a lab, collecting a representative soil sample is essential for meaningful results. Dig a few inches deep and gather small amounts of soil from 10 or more places around the garden. Mix them together to make a composite sample.

Where drainage is less than ideal, incorporating organic matter deeply into the soil will help improve it. Another option is to build raised beds for your plants. In severe cases, you'll either need to install an underground drain system or select plants that naturally thrive in wet soil.

Check Soil pH and Nutrient Levels. You can learn about the pH (acidity and alkalinity) and nutrient content of your soil by testing it at home or sending a sample to a lab. Although home tests are generally much less accurate than laboratory results, if you test repeatedly, you will soon become more adept at doing the tests and reading and interpreting the results. Getting accurate results is a key part of building healthy soil, and it will enable you to keep needed nutrients at optimum levels for plant growth.

Testing labs report soil pH and availability of major nutrients and recommend how much lime, sulfur, or fertilizer to add to improve the soil for growing whatever plants you specify. A single test can serve as a baseline appraisal of your soil. A series of tests repeated over time will monitor the effects of the treatments you apply. Keep notes on the test results, the treatments, and on how your plants look and grow, and you'll develop the eye of an experienced gardener. See "Sources" starting on page 117 in the back of this book for soil-testing labs you can use.

Soil Improvement Basics. There are several approaches to soil improvement. Tilling, hand digging, and well-timed cultivation can all help loosen and aerate soil. Installing drains or planting in raised beds improves the movement of water and air through the soil. But the real key to building and maintaining healthy soil is regular additions of organic materials.

Organic matter — in the form of compost, shredded leaves, composted bark, and similar materials — helps the soil in many ways. It increases microbial activity, which in turn improves the soil structure and makes it easier for water, air, and roots to penetrate. It also increases the soil's water- and nutrient-holding capacity, and it provides a small source of nutrients. The different forms of organic matter all have desirable effects on plant health, so choosing among them is a matter of price, availability, and convenience.

> **TIPS FOR SUCCESS**
>
> Focus soil-improvement efforts on those areas where they will have the most effect. Annual flowers and vegetables have fine roots that are too weak to penetrate hard soil, and they are healthiest when pampered with optimum conditions. Perennials thrive in improved soil, but generally have stronger, sturdier roots and can withstand less-than-ideal conditions.

Adding Organic Matter. When adding compost or other organic matter to the soil, you can add as much as 25 percent amendment to the soil in a new bed. Adding 25 percent means spreading a 3-inch layer of material on the surface and working the bed 12 inches deep (½ cubic yard of amendment per 100 square feet of bed). This will make a big difference in how the soil feels and looks and in how your plants will respond.

Changing Soil pH. A soil test will indicate the pH of your garden soil. Then you can decide whether to work with the existing condition or change it. Most plants grow well in neutral to slightly acid soil (pH 6.0 to 7.0). Chances are good that plants that are thriving in other gardens in your neighborhood are fairly tolerant of the existing soil pH. Choosing these plants is an easy place to start, but changing the soil pH somewhat in order to grow less common plants is not too difficult.

Lime is added to acid soil to raise the pH, and sulfur is added to alkaline soil to lower the pH. Both are inexpensive materials that you can purchase at local garden centers. The amount of lime or sulfur you need to add depends on the soil's texture and the amount of organic matter it contains. Check the directions on the package as well as your soil test results to determine how much material to add to make the adjustment you want. Regularly adding organic matter is another way to help neutralize overly acid or alkaline soil over time.

CHOOSE THE RIGHT PLANTS

Once you've prepared an ideal planting site, follow through by choosing plants that are well adapted to your soil and climate. They are much more likely to thrive and shrug off pest and disease problems than plants that have to struggle to survive in the existing conditions. As you go through the "wish list" of plants you want to grow, refer to your favorite gardening books to find out what conditions each plant needs to thrive. Check to see if it prefers sun, partial sun, or shade. Also look into its preferred soil conditions: does it need nutrient-rich, moist soil or a dry, sandy site, or can it adapt to a wide range of soils? The texture and drainage tests you performed when learning about your soil will help you figure out which plants are likely to succeed.

Consider the Climate. Climate also plays an important role in the health and success of your garden. By relying mostly on plants that are naturally suited to your temperatures and rainfall, you eliminate many pest and disease problems at the outset. Native plants and wildflowers indigenous to your area are usually good choices, but don't overlook plants from other areas that are adapted to similar conditions.

One of the most common ways plant adaptability is rated is by hardiness zones. These zones are based on the USDA Plant Hardiness Zone Map, which you will find on page 115. This map divides the United States and Canada into 11 zones, based on average annual minimum temperatures. Check the map to see which zone you live in, and use that information as you look through books and catalogs to help you select plants that are likely to survive in your area. A plant listed for Zones 5 to 8, for instance, could possibly grow in Zone 5, 6, 7, or 8. Zone 4 would probably be too cold, and an especially cold Zone 5 winter may be damaging. To be on the safe side, it's usually best to choose plants that are rated for one zone colder than your own.

While hardiness zones are a handy tool for judging plant adaptability, keep in mind that they are not foolproof. They do not take into account the duration of cold temperatures, snow cover, or soil drainage — all factors that can influence cold hardiness. They also don't tell you about tolerance for heat and/or humidity, which can have a great effect on plants growing in the South. Microclimates — areas on your property that offer special conditions, such as beds below warm, south-facing walls or dry, sloping sites — also influence how well certain plants will grow. So use hardiness zones as a starting point for selecting appropriate plants, but use your own observation to develop a sense of how they actually perform in your landscape.

Look for Resistance. The best way to limit the number of pests and diseases in your garden is to avoid growing plants susceptible to these problems. There are hundreds of annuals, perennials, shrubs, vines, vegetables, fruits, and trees available today, and many new releases feature resistance to pests, and especially diseases. Pick the species and cultivars that tend to have the fewest problems — ask your local nursery owner or county agricultural extension agent if you're unsure. You'll be glad you did. There are apples and crabapples resistant to a wide range of diseases, and beautiful new disease-resistant roses, for example, that reduce or eliminate the need to spray for disease problems.

If you know that a certain pest or disease is a particular problem in your area, such as black spot on roses or European corn borer on corn, pay special attention to selecting plant species and cultivars that can resist the problem. While plant breeders and growers have concentrated on selecting plants for disease resistance, there are also some pest-resistant cultivars available. Although resistant plants may not be completely damage-free, they will be much less prone to injury than other nonresistant plants growing in the same area, even without additional sprays or dusts.

If no resistant plants are available, look for those labeled "tolerant." Tolerant crops may still be attacked by pests or diseases, but they will usually still produce a usable crop without spraying.

Plan Diverse Plantings

As you plan where your new plantings will go, it makes sense to group plants that need similar conditions. But within those groupings, plant a variety of flowering and leafy plants. In the vegetable garden, for instance, you might add blocks of flowering herbs or colorful annuals between the crop plantings. By including many different species and avoiding planting all one kind of plant together, you're more likely to create a stable ecosystem with plenty of beneficial organisms that will help keep problems in check. A mixture of plants also greatly reduces the chance that your entire garden will be wiped out by a particular pest.

> **TIPS FOR SUCCESS**
>
> Get extra help controlling garden pests by attracting birds, frogs, and toads to your landscape. A small water garden will provide a great habitat for many beneficial creatures. A well-stocked bird feeder, birdhouses, and berry-producing shrubs will also attract a wide variety of feathered predators.

Attract beneficial insects. Beneficial insects can play a major role in controlling garden pests without sprays or dusts. Some beneficials act like parasites, laying their eggs in the bodies of pest insects. As the eggs hatch, the developing larvae feed on the pest. Other beneficials are predators — they attack pests directly.

Predators may control pests only during a certain period of development. For

CREATING A HEALTHY GARDEN

To attract a diverse community of beneficial insects, plant a mix of flowers and herbs. Many so-called weeds, such as Queen-Anne's lace, goldenrod, and yarrow, also attract beneficials, so consider a weedy, wild garden for them.

instance, the larvae of green lacewings are voracious predators of aphids and other soft-bodied insects. Once they mature, however, the adults rely on pollen and nectar for food. Other beneficials may feed on pests throughout their lives, but they still need pollen and nectar for supplemental food. For these reasons, it's important to include plenty of nectar- and pollen-producing plants in your garden. Small-flowered plants are particularly good; try some of the following:

- Carrot-family plants, such as caraway, dill, fennel, and parsley
- Daisy-family plants, such as purple or orange coneflowers *(Echinacea* or *Rudbeckia* spp.), Shasta daisies *(Chrysanthemum* × *morifolium),* and yarrow *(Achillea* spp.)
- Mint-family plants, including catnip, mints, oregano, rosemary, sage, and thyme

Mixed plantings encourage beneficial insects such as lacewings. The adults eat pollen and nectar from many herbs as well as daisy-family plants. Their larvae consume aphids and other soft-bodied insects.

Spiders, such as this golden garden spider, catch and eat a wide variety of insects and are welcome garden residents.

Besides providing food sources, you can create ideal conditions for beneficials by supplying water and shelter. During dry periods, make water available by filling a shallow pan with rocks and adding water to almost cover the rocks. This will allow the insects to land and drink without falling in. Provide shelter with ground covers, shrubs, perennial borders, and other areas where the soil and plants are seldom disturbed. You can also help beneficials by limiting the use of *all* sprays and dusts. Even pesticides that are considered acceptable to organic gardeners may harm beneficial insects if used carelessly.

Discourage disease development. Crop rotation can be as important for preventing disease problems as beneficial insects are in controlling garden pests. Crop rotation simply means shifting plantings to different sites every year so that you aren't growing the same plants in the same place each year. Obviously, it's mainly useful for annual flowers and vegetables — things that get planted each year. But it can also be important when replacing other plants. Roses, for instance, often suffer if planted where another rose has recently died.

Crop rotation helps by preventing the buildup of any one kind of disease organism in the soil. Many diseases will only attack a few kinds of related crops. So when you grow different crops each year in a given site, the pathogens will eventually die out from lack of food.

The simplest rotation system is to just grow each kind of plant in a different spot each year. For instance, if you planted petunias by the front door last year, you could grow marigolds there this year and plant the petunias somewhere else.

To get even more good out of crop rotation, you need to take into account the botanical family each plant belongs to. The cabbage family, for example, includes cabbages, cauliflower, broccoli, and several other common crops. All of these cabbage relatives are prone to a disease called club root, which causes distorted roots and poor growth. To minimize club root problems, you should make sure that you don't grow one cabbage-family crop after another (cabbage after broccoli, for example) in the same location.

To help you plan the most effective rotations, here's a quick summary of the major garden plant families and their members:

- Beet family: beet, Swiss chard
- Cabbage family: Brussels sprouts, cabbage, cauliflower, collard, kale, kohlrabi, radish, sweet alyssum, turnip
- Carrot family: carrot, celery, dill, fennel, parsley, parsnip
- Daisy family: ageratum, calendula, cosmos, lettuce, marigold
- Grass family: corn
- Onion family: chives, garlic, onion, shallot
- Pea family: bean, pea, sweet pea
- Squash family: cucumber, melon, squash
- Tomato family: eggplant, flowering tobacco, pepper, petunia, potato, tomato

Where garden space is limited, you may not have much freedom to move plants to different sites each year. Whenever possible, though, try to leave three to six years between related crops. If you really want to grow the same crop in the same place each year, at least try a different cultivar.

BUY HEALTHY PLANTS

Once you've created the ideal planting site and selected suitable plants, the last thing you want to do is bring in pests, diseases, or weeds on the plants you buy. To minimize the chances of this happening, always buy from reputable sources and inspect purchased plants carefully before you bring them home.

When shopping, choose nurseries and garden centers that look clean and well maintained. Weeds and debris lying around can host pests and diseases, not to mention weed seeds. The plants themselves should be healthy looking, with evenly colored leaves and stems, and the pots should be weed-free. Avoid those that are wilted or off-color. (Plants left to fry on a hot sidewalk outside a dis-

---

Starting Healthy Seedlings

Raising your own transplants from seed is one way to make sure you're getting healthy, well-grown seedlings. Give your young plants a good start by sowing them in a sterilized medium, such as commercial seed-starting mix. Don't use garden soil for indoor seed-starting; it may contain insects and diseases, as well as weed seeds.

The major problem seedlings face is a disease called damping-off. They may be killed even before they emerge, up to when they are an inch or two tall. Infected seedlings develop a dark or water-soaked patch at ground level and simply topple over. Damping-off can spread quickly through a whole pot or tray of seedlings.

Besides using sterilized seed-starting medium, you can discourage damping-off by providing good air circulation around the seedlings. Place a small fan, set on low, a few feet away from your seed pots. Set the fan to stir the air just above the surface of the pots until seedlings are a few inches tall. Sprinkling finely ground sphagnum moss over the surface of the pots just after sowing can also help prevent disease problems.

count store may be inexpensive, but they are often stressed and unhealthy because of erratic watering by inexperienced store personnel.)

Also check to see if the plant looks in proportion to its pot. If possible, slide the plant out of the container and look at the soil ball. If you see many roots that are matted or circling around the soil ball, the plant is overgrown, and it may have a hard time adjusting to your garden. Lastly, look closely at the stems and leaves (including the undersides). Check for tiny white flying insects (whiteflies), unusual smooth or fuzzy bumps on stems or leaves (scale insects or mealybugs), and small pear-shaped insects clustered on leaves, stems, or buds (aphids). Also avoid any plants with unusually streaked or spotted stems or leaves.

There may be times when generous friends or neighbors will give you a gift plant, or perhaps some divisions from their own plants. Before adding these gifts to your garden, give them the same going-over you would if you were buying them. Treat any pest or disease problems, if possible, or discard the plant discreetly if it is seriously diseased or infested. It's a good idea to set up a "quarantine bed" in a corner of your landscape, where you can grow new acquisitions for at least a few weeks before moving them next to healthy plants.

Provide Proper Plant Care

Many routine gardening activities, like mulching, fertilizing, and clearing away debris, also play a role in fighting pest and disease problems. You may not think you are preventing pest problems when you mulch perennials or feed your vegetable garden, but you are. By providing the best possible care, you're helping your plants to be naturally more resistant to all kinds of problems.

Mulch management. Using mulch effectively is a cornerstone of good plant care. Organic mulches, such as chopped leaves, shredded bark, or compost, help to keep the soil cool and moist, providing ideal conditions for good root growth. They prevent rain and irrigation water from splashing soil and disease-causing organisms onto leaves and stems. By shading the soil, mulches can also prevent many weed seeds from germinating. And as they break down over time, organic mulches release nutrients and humus to improve the soil structure and promote healthy plant growth.

Plant properly

Transplanting is stressful for any plant you add to your garden. You can prevent problems and get plants off to a healthy start by preparing the soil properly, handling them carefully, and watering regularly until growth resumes.

It pays to know proper planting procedures for each type of plant. When planting a clematis, shown here, set the top of the root ball 2 to 3 inches below the soil surface.

To get the most from your mulch, apply it to soil that has already been weeded; it won't suppress weeds that have already sprouted. Use a 2- to 3-inch layer of mulch, and add more as needed several times a year to maintain the right thickness. Don't pile mulch against plant stems, as it can hold in moisture and encourage rot. Leave a mulch-free zone about 1 inch wide around annuals, perennials, and vegetables; leave a 6- to 12-inch, unmulched circle around trees and shrubs.

In damp areas, you may have a problem with slugs and snails making a home in the mulch. If you notice damage on mulched plants, remove the mulch until the weather dries out. If your area tends to be damp year-round, you may need to use just a ½- to 1-inch layer, or perhaps avoid using mulch altogether.

Watering wisely. By improving your soil, choosing adapted plants, and using mulch, you'll dramatically reduce the amount of watering you have to do every year. But it's likely that you will still have to add supplemental water at some point to get top yields from a certain crop or help plants through a dry spell.

The most effective way to get water to thirsty plants is with a soaker hose or a drip or trickle irrigation system. These pipes, laid along rows or snaked around plants growing in a bed, slowly release water right onto the soil. This way, the roots get the moisture they need without the plants getting wet. (Moisture sitting on plant leaves provides prime conditions for disease development.)

The key to effective irrigation is watering deeply but infrequently. Each time you irrigate, let the water run until the top 4 to 6 inches of soil are moist. Then wait until the top 3 or 4 inches are dry before watering again. This will encourage plants to develop deep root systems, so they'll be better able to withstand dry spells without wilting.

Fertilize for healthy growth. Supplying a balance of soil nutrients is another key part of good plant care. Earlier in this chapter, you learned about the importance of soil tests for identifying and correcting nutrient imbalances. Working compost or other organic matter into the soil before planting will also provide a small but steady supply of nutrients over time. Testing new soil samples every year or two will help you identify and correct nutrient deficiencies before they affect the health and vigor of your plants.

A wide variety of organic fertilizers are available through garden centers and garden supply catalogs. Follow the recommendations on your soil test results or use the application rates on the package to determine how much to apply. Don't be tempted to add more than you are supposed to; too much of a particular nutrient can be as harmful to plants as too little. Scatter dry fertilizers on the soil and scratch them lightly into the surface. Liquid fertilizers, sprayed directly on leaves or watered into the soil, are useful for providing a quick nutrient boost during the growing season.

Pruning and grooming. A little regular trimming and pinching can work wonders to keep your plants looking good and growing vigorously. Thinning out crowded stems will allow good air movement through a clump, helping wet leaves

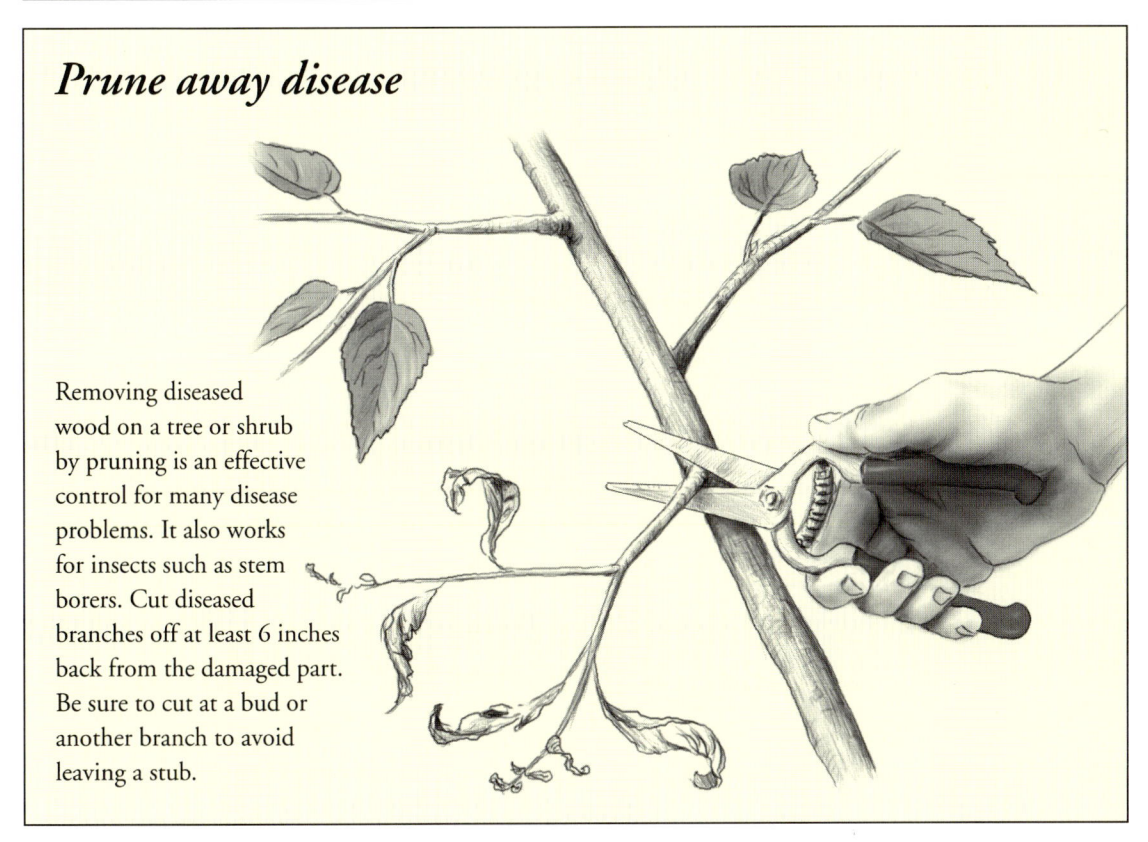

> ## Prune away disease
>
> Removing diseased wood on a tree or shrub by pruning is an effective control for many disease problems. It also works for insects such as stem borers. Cut diseased branches off at least 6 inches back from the damaged part. Be sure to cut at a bud or another branch to avoid leaving a stub.

and stems dry faster and discouraging disease development. You can get the same benefits by dividing overgrown perennial clumps in spring or fall.

During the growing season, pinch off and destroy insect-infested tips or discolored leaves as soon as you spot them. This simple step can often stop developing problems before they spread to other plants. You can remove diseased or dead wood from trees and shrubs any time of year, although it's best to choose a dry day. Always cut back to a main stem or to an outward-facing bud; stubs can provide easy entrance for pests and diseases.

Keep the garden clean. Good sanitation is a vital part of reducing problems with insects, mites, nematodes, and diseases. During the growing season, eradicate weeds near your garden. Besides competing with your desired plants for available light, water, and nutrients, weeds often harbor insects and diseases. Regularly pick up dropped leaves and fruit. Also, destroy (don't compost) diseased or infested plants and plant parts.

At the end of the growing season, cut down and rake up withered leaves, stems, and fruits; add them to your compost pile. Unless you are saving them for winter interest, remove seedheads from herbs and ornamental plants to prevent them from producing unwanted seedlings all over your garden. In the vegetable garden and annual beds, cultivate the soil often to expose insect eggs and larvae and disease-causing organisms — sunlight and wind can desiccate and kill many of them.

Keep Garden Records

Sometimes, casual observation of your plants will give you all the information you need to judge their health. Other times, more serious and regular inspection is in order. In either case, you may find it helpful to carry a notebook with you and record damage or anything else unusual that you observe on your plants. Also record the date you first notice the problem and how it changes over time.

Use your notebook to keep track of your garden activities, such as fertilizer applications or crop rotations, as well as the environmental conditions. Outbreaks of many pests and pathogens are triggered by changing weather conditions. A prolonged hot, dry spell, for example, often will be followed by an increase in spider mites; weeks of rainy weather may encourage fungal outbreaks. This record-keeping may seem like a lot of trouble, but the information will provide invaluable clues for tracking down the causes of problems.

Scout for Problems

Check your garden regularly for damaged plants, as well as insects and other organisms. Over time, careful observation will help you recognize when something's gone awry and will give you a better sense of local pest and disease cycles.

Keep an eye out for damage. Take an up-close look: Check the top and underside of the leaves, along the stems, where the branches meet the stem, buds and flowers, the base of the plant and the surrounding soil surface. Note leaves that are munched, covered with webbing, or abnormal in shape or color. Pay attention to whether the damage occurs along the leaf margin, the veins, or throughout the leaf. Also note whether holes or discolored areas are regular or irregular in shape. Look for wilted and blackened or chewed stems, sticky exudate on branches, withered or browned flower buds, or loose mounds of soil near the plant.

Look for insects — a 10× hand lens will expose tiny ones. Check for egg masses on or near the plants. Also take a look at plants in nearby wild areas, where pest species may find a habitat to their liking. Occasional nighttime forays into the garden with a flashlight might reveal critters that are hidden from view in the day.

Identify the cause of the problem. Once you see damage to a plant, the first task is to identify the culprit. This is easier said than done, but essential if you want to solve the problem with the safest, most effective techniques. Misidentifying the pest or disease that caused the problem is a common occurrence. Many insects look similar to the untrained eye, and sometimes the perpetrator has left the scene entirely before you arrive. Pests also sometimes hide within the plant

Nasty Nematodes

Nematodes are microscopic roundworms, some species of which attack roots, while others attack leaves. Beneficial nematodes help decay organic matter in the soil. Some lso are predatory and help control grubs and other root pests. Problems caused by pest nematodes can be especially difficult to identify, since the symptoms they produce can resemble many other types of damage. Infested plants may be stunted or off-color, and they don't improve with extra water or fertilizer. Other symptoms include root galls, excessive root branching, and/or galled or distorted leaves.

The best ways to foil nematodes are to plant nematode-resistant selections, add organic matter to the soil, and rotate susceptible crops every year. You may also want to try Clandosan, a recently introduced natural control, or beneficial predatory nematodes. Don't move plants from infested areas to clean soil. If you do, you'll spread the pests.

or operate underground. It's all too easy to conclude that whatever insect is present at that time is the culprit. Launching an all-out attack on a benign or even beneficial insect that just happens to be in the wrong place at the wrong time won't solve the problem and may even make it worse.

One of the best clues to the identity of a pest is the damage it causes. Chewing insects, such as beetles and caterpillars, eat holes in leaves or munch away surface leaf tissue, leaving a sort of leaf "skeleton." Cutworms often nip off seedlings at ground level. White, irregular shaped pathways on the surface of a leaf mark the trails of leafminers, which tunnel their way between the layers of leaf tissue. Borers burrow into stems and branches, sometimes leaving oozing, sticky sap or sticky sawdust in their wake. Slugs and snails also chomp leaves and leave slimy, shiny trails.

Insects such as aphids, mealybugs, scale, and whiteflies, as well as mites (which are more closely related to spiders than insects), all suck the sap from plants. This disruption of normal plant growth causes a variety of symptoms, ranging from wilted, twisted, yellowed, spotted, or curled leaves to stunted or dead plant parts. If aphids or scale are causing the damage, there will often be an accumulation of honeydew, a sticky shiny substance, on leaf surfaces as well. Honeydew is often coated with black sooty mold, a fungal disease.

Plant diseases are most often caused by fungi, bacteria, or viruses. But pathogens are more elusive than insect pests — most are microscopic — so your best bet is to try to identify them by their typical symptoms or signs. While these symptoms are generalizations at best, they should point you in the right direction.

- Mildews are recognized by a gray-white powdery coating on leaf surfaces, stems, or fruit, sometimes with yellowed or dead areas beneath.
- Leaf spots, which can result from fungi or bacteria, appear as discolored lesions on the leaves — yellow, red, brown, tan, gray, or black spots. Sometimes the infected leaves wilt and rot.
- Fungal or bacterial blights cause leaves, branches, twigs, or flowers to suddenly wilt or become brown and die. Often, the stem appears water-soaked and blackens at the soil line.
- Rots, caused by fungi or bacteria, can cause similar decay of roots, of the lower part of the stem, or of other succulent plant tissue.

The white, ricelike pupae on this tomato hornworm make it more of a garden friend than foe. The pupae indicate the hornworm has been parasitized by braconid wasps. When it dies, the wasps will emerge and parasitize a new host. Braconid wasps also attack cabbageworms, corn borers, armyworms, and other insects.

- Viruses can produce white, yellow, or pale green discolored, patterned, or spotted areas on leaves, and less commonly on stems, fruit, or roots. Virus-infected plants often become stunted.

For help in identifying the insects, other organisms, or symptoms you encounter, refer to "A Guide to Garden Pests," starting on page 35 and "A Guide to Garden Diseases," starting on page 68. Reliable diagnosis of plant problems is often a tricky business, so you shouldn't feel reluctant to turn to an expert for assistance. Your cooperative extension service (listed under city or county government offices in the phone book) or a local nursery or garden center may be able to help or refer you to someone who can.

Keep in mind that not every problem is caused by an insect or other pest organism — damage also can be caused by cold, heat, wind, air pollution, too much or too little water or fertilizer, or other environmental and cultural factors. To make matters more confusing, sometimes this damage looks similar to that caused by pests.

CHOOSE THE RIGHT CONTROLS

When you spot a potential problem and identify the culprit, think before you take action. Learn as much as you can about the insect's or disease's life cycle and preferred environment. Monitor pest populations to see if their numbers are increasing, decreasing, or holding steady. Check for insects or other organisms that prey on the pest; given time, they may provide adequate control. Remember, populations of beneficial organisms are often slower to build up than pest populations — taking action too soon could wipe out the good guys. Even an ill-timed spray of an apparently harmless substance like water can wash off beneficial predators.

Decide if you and your plants can live with the damage. Specific guidelines have been developed to help large-scale growers of commercial crops figure out just how many of a particular insect pest are too many, but home gardeners are left to their own judgment. Keep in mind that beneficial insects need a supply of pests to maintain their populations. If a pest-covered plant is seriously damaged and going downhill fast, and there are no natural enemies in sight, you need to think about taking action. On the other hand, if you spot a few pests here and there, but they aren't causing much harm, and they aren't on the increase, then you might take a "wait and see" attitude. You'll also need to think about your aesthetic standards and decide just how much damage is acceptable to you.

If you choose to take action, start with the most benign controls, ones that you can carry out with just your hands (or feet), and move onto other tactics as needed. In the end, a combination of strategies is often most successful.

Cultural controls. First, change any cultural practices that might make your plants more susceptible to a particular pest or disease. For example, planting earlier might give your plants a chance to establish themselves before the organisms that attack them are up and running. Another strategy is to modify a pest's habitat in a way that discourages its survival. For instance, you might remove a thick mulch or clear out adjacent weedy areas where the pests hide when they're not feeding in your garden.

Mechanical and physical controls. Traps, barriers, cages, row covers, and other mechanical devices or barriers can exclude pests before they arrive and create a pest-free zone once you've rid plants of existing pests. Physical controls, such as handpicking pests or crushing them underfoot, are particularly successful for larger critters such as beetles, snails, and slugs. Strong water sprays can physically dislodge certain insects. All of the above methods of control pose no risk to people or the environment.

Biological controls. Increasingly, gardeners are making use of their pest's naturally occurring enemies: predators, which are free-living organisms that feed on other organisms; parasitoids, which kill the hosts they live on; and pathogens, which are microorganisms that release toxins into the insects that ingest them. These beneficial insects, mites, nematodes, and microorganisms are probably already at work naturally in your garden. Under the best of circumstances, you can just sit back and enjoy the fruits of their labor.

If, however, these beneficial organisms haven't appeared yet or are in short supply, you can purchase laboratory-reared populations to release in your garden. There are commercially available beneficials that control pest insects, mites, and snails. Some, like the larvae of the green lacewing, can be quite effective, while others, like the convergent lady beetle, often fly away before they've adequately

Inexpensive physical controls to make yourself

CARDBOARD COLLAR
A simple cardboard circle will deter adult cabbage maggots from laying eggs at the base of cabbage seedlings. Cut a slit in one side of the circle, with a small hole for the stem to slip through. Slip it around the seedling at planting time so the collar lies flat on the soil.

CUTWORM COLLAR
To prevent soil-dwelling cutworms from chewing through the stems of newly planted seedlings, slip a small soup can (both ends removed) or half a toilet paper tube over each seedling. Push the collar halfway into the soil.

Apple Maggot Trap

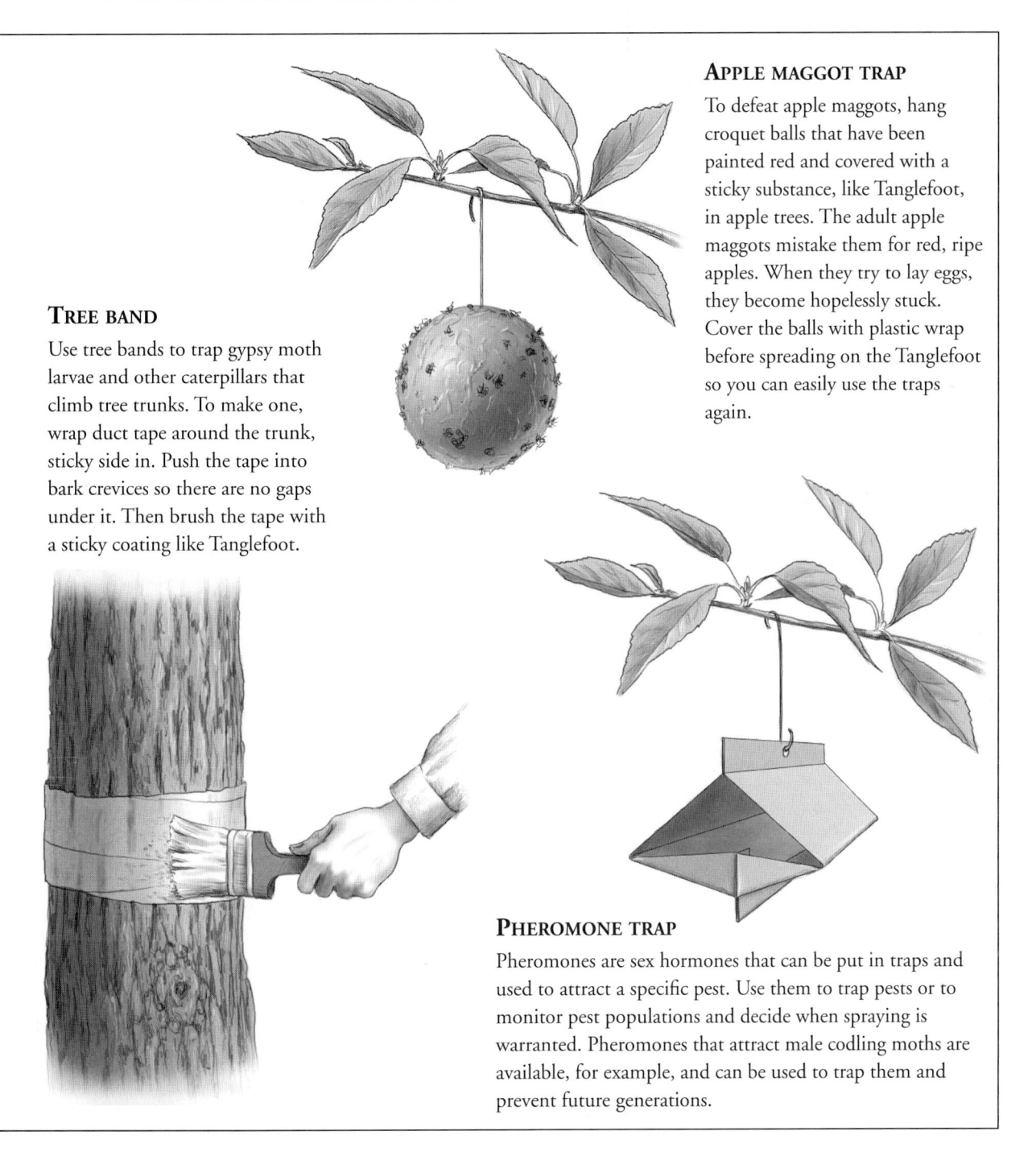

To defeat apple maggots, hang croquet balls that have been painted red and covered with a sticky substance, like Tanglefoot, in apple trees. The adult apple maggots mistake them for red, ripe apples. When they try to lay eggs, they become hopelessly stuck. Cover the balls with plastic wrap before spreading on the Tanglefoot so you can easily use the traps again.

Tree Band

Use tree bands to trap gypsy moth larvae and other caterpillars that climb tree trunks. To make one, wrap duct tape around the trunk, sticky side in. Push the tape into bark crevices so there are no gaps under it. Then brush the tape with a sticky coating like Tanglefoot.

Pheromone Trap

Pheromones are sex hormones that can be put in traps and used to attract a specific pest. Use them to trap pests or to monitor pest populations and decide when spraying is warranted. Pheromones that attract male codling moths are available, for example, and can be used to trap them and prevent future generations.

Floating row covers are a very effective way to keep aphids, leafhoppers, and other pests off plants entirely. Cover rows at planting time and tuck the edges in with soil to prevent pests from crawling under them. For plants that need cross pollination, such as melons, remove the covers when flowers appear.

controlled their prey. Sometimes the introduced beneficials will establish large enough populations in your garden to provide ongoing control; in other cases you'll need to replenish the supply periodically.

Microbial insecticides are commercially available as well. These pathogens include bacteria, fungi, and viruses that are effective against insects and in some cases, against other plant diseases. One of the most common microbial pesticides,

Egg cases of praying mantids are commonly available for sale, but while these insects are effective predators, they consume pest and beneficial insects alike. Upon hatching, the young mantids also consume each other.

Bacillus thuringiensis (commonly called BT), is a toxin-producing bacteria that kills caterpillars and insect larvae after these pests have eaten treated leaves or stems. BT breaks down quickly in sunlight and is heat-sensitive, so it may require reapplication and must be stored in cool conditions. Depending on the particular pest you want to control, you can choose among several varieties of BT.

Lady beetles are familiar garden residents that voraciously consume aphids, mealybugs, spider mites, and soft-bodied scale insects.

Organically acceptable sprays and dusts. A wide range of compounds — including soaps, oils, diatomaceous earth, botanical pesticides, and minerals — are sprayed or dusted on plants to control pests and diseases. Made of fatty acids, insecticidal soaps penetrate the coating of susceptible insects and dissolve their cell membranes, causing death. They are most effective against soft-bodied insects with sucking mouthparts, such as aphids, mites, whiteflies, and thrips. Fungicidal soap formulations, made of sulfur and fatty acids, are used to control diseases such as mildews, rots, leaf spots, and rust. These soaps, sold as liquid formulations, are sprayed on affected plants. They are generally very safe to use and break down rapidly in the soil, but they can be poisonous to plants (phytotoxic). Test

This ferocious-looking insect isn't a pest at all. It's a lady beetle larva. Identifying pests before trying to control them will help you avoid harming your allies.

them on a small portion of a plant and wait a day or two to make sure no damage is evident before spraying the whole thing. Thorough coverage and periodic reapplication are needed for good control.

Horticultural oils are employed to smother the adults and eggs of a variety of pests, including certain aphids, scale, caterpillars, and mites. Traditionally sprayed when the plant is dormant, newer formulations are safe for spraying on leafed-out plants as well. Follow label precautions to avoid potential phytotoxicity, and wear protective clothing to prevent eye and skin irritation during application. The oils have a low toxicity to humans and wildlife and biodegrade rapidly.

Diatomaceous earth is an abrasive dust composed of the skeletal remains of microscopic marine creatures. It is sometimes sprinkled on the ground as a barrier to protect plants from snails and slugs — with varying results. Diatomaceous earth is nontoxic to mammals, but can irritate eyes and lungs, so wear goggles and a dust mask when you're applying it.

Botanical pesticides are derived from plants that contain substances toxic to insects. Some botanicals, such as neem and pyrethrins, are effective against a wide variety of pests (they're called broad-spectrum pesticides); others, such as sabadilla, kill a more limited range of pests. Botanicals also vary widely in their toxicity to humans, other mammals, and insects. Regardless of their toxicity, botanicals tend to break down quickly into nontoxic compounds in the presence of sunlight or

Safe Spraying

When used correctly, organic sprays and dusts can greatly benefit the home gardener. To ensure you use these materials properly, safely, and effectively, follow these guidelines.

- Always read the label. The label tells you what pests a particular product controls, how much to use, when to spray, and what precautions to take. It is also important to always follow label directions for storage and disposal.

- Know what problem you're trying to control. Fungicides won't kill insects and insecticides won't kill fungi. Using the wrong control wastes money and time and can make the problem worse. Don't expect a single product to control everything.

- Apply sprays and dusts only when there is a problem. Besides injuring beneficial creatures, overuse of pesticides can lead to resistant, "super" insects and fungi that are nearly impossible to control. So spray only when you see a pest, not before.

- Don't spray flowers. Doing so may kill honeybees, which pollinate countless trees, shrubs, vegetables, and flowers. Spray in early morning or late afternoon, when bees aren't active.

- Wear protective clothing when spraying. This includes a long-sleeved shirt and long pants, rubber gloves, boots, goggles, and a mask over your nose and mouth.

When you notice a potential pest problem, always check for beneficial insects like this lacewing, which may already be at work controlling the pests. In fact, a few pest insects actually are necessary so your garden can feed a healthy number of beneficials.

in the soil. They may require frequent reapplication, but their short period of toxicity minimizes potential harm to organisms other than the targeted pests.

Sulfur, copper, and lime — applied as dusts or mixed with water — are primarily used to control fungal and bacterial diseases, including mildews, rots, leaf spots, and blights. Sometimes, these minerals are sold in a combined formulation, such as copper sulfate, or bordeaux mix, a mix of copper sulfate and hydrated lime. Sulfur dust, lime, and lime-sulfur control some plant-sucking pests as well. Sulfur is relatively nontoxic to humans; the toxicity of copper varies depending on the formulation. Whether you apply minerals as a dust or a spray, protect your eyes, lungs, and skin to avoid irritation.

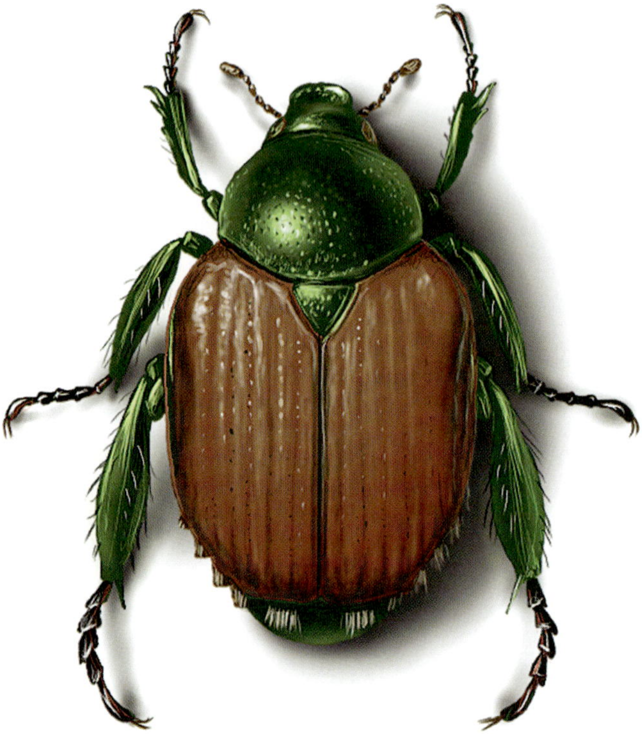

Chapter 2:
A Guide to Garden Pests

Identifying pests correctly is a key part of choosing and using the most effective control measure. If you notice pests or their damage on your plants, skim through the following pages to find out what's causing the problem and what to do about it. You will find descriptions of the pests along with information on their life cycles. In many cases, pests are easiest to control at a specific stage. In addition, controls are listed for each pest. Preventive techniques are listed first, followed by "soft" controls and stronger, plant-derived organic sprays.

Keep in mind that having a few pests is no problem for most plants in most situations. When you look at a plant that seems to have a pest problem, try to decide if the pests are really damaging the plant to the point that a problem may occur. Most plants easily can withstand a few chewed leaves or petals, for example. If, for some specific reason, you require perfect foliage or a perfect flower, treat those plants separately and as needed.

Many beetle larvae chew on plant roots. Applying parasitic nematodes to the soil is an effective, long-term control for these pests.

Aphid (actual size)

- **Aphids**

DESCRIPTION: Adults and nymphs are tiny green, black, brown, or reddish, pear-shaped, soft-bodied insects. Some have wings. Look for these pests in clusters on buds, shoots, and undersides of leaves.

DAMAGE: Aphids suck plant juices, causing stunted or deformed blooms and leaves. As they feed, they also exude a sticky substance called honeydew, which produces a shiny coating on leaves and supports the growth of black sooty mold fungus. Aphids may also transmit plant viruses. They attack a wide range of vegetables, herbs, flowers, fruits, shrubs, and trees.

CONTROL: Encourage or introduce natural predators, including lacewings and lady beetles. Pinch off and destroy infested plant parts, crush aphids with your fingers, or knock the pests off plants with a strong spray of water. Spray serious infestations with insecticidal soap or pyrethrins.

Apple maggot larva damage

- **Apple maggots**

DESCRIPTION: Larvae are 1/4-inch whitish or yellowish maggots. Adults are slightly smaller than houseflies, with a black body, a yellow-striped abdomen, and yellow legs. The transparent wings are marked with four dark bands.

DAMAGE: Adult females lay eggs under apple skins in early to midsummer. The eggs hatch into larvae that tunnel through the fruit of apples, blueberries, cherries, crab apples, pears, and plums. Infested fruit drops early.

CONTROL: Maggots can't be controlled once they are in the fruit. Pick up and destroy dropped fruit daily (or at least weekly) through the summer; this will prevent larvae from leaving the fruit and overwintering in the ground. Reduce egg-laying by trapping adult flies. Buy commercial traps or make your own by covering an apple-sized red ball with plastic wrap. Coat the plastic wrap with a sticky substance, such as Tanglefoot (available in garden centers). In early summer and midsummer, hang one trap in each small tree, or up to four in each full-size tree. Replace the wrap and sticky coating every two weeks.

Armyworm (actual size)

■ Armyworms

DESCRIPTION: Armyworm larvae are pale green to greenish brown caterpillars with stripes on their sides and backs. Adults are pale gray-brown moths.

DAMAGE: Larvae feed only at night, devouring leaves and even whole plants. During the day, they hide under garden debris. Armyworms attack many different plants, including vegetables and lawn grasses.

CONTROL: Handpick pests at night, or spray damaged plants with BTK *(Bacillus thuringiensis* var. *kurstaki)* or neem. Prevent damage by covering most crops with floating row cover from planting time until harvest. (To allow for pollination, remove row covers from squash, cucumbers, melons, and other insect-pollinated crops as soon as flowers appear.) Grow plants that are rich in pollen and nectar to attract armyworm predators.

Bagworm
(actual size)

■ Bagworms

DESCRIPTION: Bagworms are the larvae of moths. You'll rarely see these dark brown caterpillars, however, since they hide inside a spindle-shaped cocoon or "bag" that they carry along as they feed. These cocoons are up to several inches long and are covered with bits of leaves or needles from the host tree. Male adults are black moths with clear wings; adult females are wingless and remain in the cocoon.

DAMAGE: Bagworms chew on the foliage of many evergreens and deciduous trees through the summer. Severe infestations may defoliate portions of plants. The cocoons may also be unsightly.

CONTROL: Spraying with BTK *(Bacillus thuringiensis* var. *kurstaki)* in early spring may help to control young larvae. To minimize damage to next year's growth, handpick and destroy bags during the winter. Use a knife to cut the cocoons from the twigs. Grow pollen- and nectar-rich flowers to attract bagworm predators.

■ Beetles

DESCRIPTION: Beetles are hard-shelled oval to oblong insects. While certain kinds of beetles are beneficial in the garden, others are troublesome pests. Among the most common pest species are Japanese beetles, asparagus beetles, and rose chafers. Adult Japanese beetles are metallic-looking green insects with coppery brown wings; the larvae are brown-headed white grubs. Asparagus beetles are reddish brown and black with four white blocks on their wing covers; the larvae are greenish gray grubs with black heads. Rose chafers, also known as rose bugs or rose beetles, have reddish brown bodies and long legs; the larvae are small white grubs. For information on other common beetles, see "Colorado Potato Beetles" on page 43, "Cucumber Beetles" on page 45, "Flea Beetles" on page 46, and "Mexican Bean Beetles" on page 54.

Japanese beetle (actual size)

DAMAGE: Pest beetles chew holes in leaves, stems, and flowers during the growing season. The larvae of some kinds feed on roots. Adult Japanese beetles and rose chafers attack many different garden plants, while asparagus beetles are limited to asparagus. Japanese beetle grubs are a pest of lawns, producing patches of dead or wilted grass. (For more details, see "Grubs" on page 49.)

CONTROL: Handpick adult beetles early in the morning and drop them into a container of soapy water. Apply parasitic nematodes to the soil to control grubs. You can also apply spores of milky disease to lawns to control Japanese beetle grubs. Treat seriously infested plants with neem, pyrethrins, or rotenone.

Peach borer
(actual size)

■ Borers

DESCRIPTION: Borers are the wormlike larvae of moths or beetles. These larvae tunnel into leaves, stalks, branches, or trunks and feed inside the plant. A few of the most common kinds include appletree borers, corn borers, fruit borers, iris borers, peachtree borers, squash borers, and stalk borers. You'll seldom see the borers themselves, so it's important to be familiar with the symptoms they produce.

DAMAGE: Adult borers lay eggs in or on stems and leafstalks. The larvae burrow into leaves, stems, or roots. Look for small holes — often close to the ground — with gummy, sawdustlike material around or just below them. Affected plants are weakened or killed. The various kinds of borers attack many different plants, including fruit trees and bushes, ornamental trees and shrubs, flowers, and vegetable crops. Wilting even when the soil is moist is a common symptom on flowering plants and vegetable crops infested with borers. Corn borers may feed on tassels and ears on corn plants and tunnel into the stems as well. A slimy bacterial rot often follows iris borer feeding in iris rhizomes.

CONTROL: If possible, dig out and destroy borers. On trees and shrubs, inserting a flexible wire into the entrance hole may impale and kill the borer. Injecting parasitic nematodes into borer holes may also help. Dig up iris rhizomes and destroy infested leaves and stems. Avoid damaging tree and shrub stems with lawn-care equipment; wounds provide easy access for borers. Cover squash, cucumbers, melons, and related plants with floating row covers from planting to flowering to minimize borer damage.

A Guide to Garden Pests

■ Caterpillars

DESCRIPTION: Caterpillars are soft-bodied, wormlike creatures with several pairs of legs. Their bodies may be smooth, hairy, or spiny. Adults are moths or butterflies. Many different kinds of caterpillars attack garden plants. A few of the most common include cabbage loopers, cankerworms, corn earworms, imported cabbageworms, and tomato hornworms. Cabbage loopers are green and 1 to 2 inches long. Cankerworms are about 1 inch long, with white stripes along their greenish, yellowish, or brownish bodies. Both cabbage loopers and cankerworms "loop" their bodies as they move. Corn earworms are 1 to 2 inches long, with yellow, green, or brownish, striped bodies. Imported cabbageworms are light green and about 1 inch long. Tomato hornworms are thick, white-marked, green caterpillars between 4 and 5 inches long.

Cankerworm (actual size)

Several other kinds of caterpillars cause similar damage. For information on other leaf-feeding caterpillars, see the following entries: "Armyworms" on page 37, "Bagworms" on page 38, "Conifer Sawflies" on page 44, "Cutworms" on page 45, "Gypsy Moths" on page 50, "Leafrollers" on page 53, "Sod Webworms" on page 62, and "Tent Caterpillars" on page 64.

DAMAGE: Caterpillars attack many different plants, chewing holes in leaves, flowers, fruits, and shoots through the growing season. Cabbage loopers and imported cabbageworms are serious pests of cabbage-family plants. Cankerworms feed on trees and shrubs. Corn earworms damage corn silks and ears and tomato leaves and fruits, as well as other vegetables and flowers. Tomato hornworms chew on tomatoes, potatoes, and related crops.

CONTROL: Handpick and destroy caterpillars. (Wear gloves to prevent possible skin irritation.) Spray leaves and fruit with BTK *(Bacillus thuringiensis* var.

kurstaki) or apply granular BTK to corn ears when silks start to dry. Sticky tree bands will catch cankerworms. Pyrethrins or neem may help control serious caterpillar infestations. Prevent damage on vegetable crops by protecting them with floating row covers.

Codling moth larva and damage

▪ Codling moths

DESCRIPTION: Codling moth larvae are ¾-inch pinkish white caterpillars with brown heads. Adults are grayish brown moths with wavy brown lines on the forewings.

DAMAGE: Codling moths emerge and lay eggs around the time the flower petals drop; larvae hatch and begin burrowing into fruit in one to three weeks. They cause blemishes on the skin with crumbly brown excrement. Codling moth larvae tunnel through the fruits of apples, pears, and quince. They leave the fruit to overwinter under bark or in leaf litter under the tree.

CONTROL: Once larvae are in the fruit, there's not much you can do to save that crop. To minimize damage to future crops, use sticky tree bands to trap the larvae that are leaving the tree to pupate. In late winter, scrape loose bark off trees to expose any overwintering cocoons, and apply a dormant oil spray. Choose resistant cultivars when planting new trees.

- **Colorado potato beetles**

    DESCRIPTION: Adult beetles are orange with black and creamy white stripes on their wing covers. The larvae are plump orange grubs with black dots along the sides. The yellow eggs are laid in masses on the undersides of leaves.

    DAMAGE: Larvae and adults chew holes in the leaves and stems of tomatoes, potatoes, eggplants, petunias, and other tomato-family crops.

    CONTROL: Handpick adults and larvae and drop them into a container of soapy water. Crush egg masses with your fingers. Attract predators and parasites with pollen- and nectar-producing plants. Use a deep straw mulch around plants, or cover them with floating row covers. To control serious infestations, spray leaves weekly with BTSD (*Bacillus thuringiensis* var. *san diego*), neem, or pyrethrins.

Colorado potato beetle (actual size)

Conifer sawfly larva (actual size)

- **Conifer sawflies**

    DESCRIPTION: Conifer sawfly larvae resemble caterpillars, with grayish, 1-inch bodies and dark heads. Adults are thick-bodied wasps that are less than 1/2 inch long.

    DAMAGE: They may eat entire needles or just the edges; severe infestations can defoliate trees. Feeding begins in early spring. Larvae drop to the ground in late spring or early summer and pupate in cocoons in leaf litter. Some species overwinter in these cocoons; others emerge as adults in fall and lay eggs in needles. Conifer sawfly larvae cause serious damage on hemlocks (*Tsuga* spp.), pines (*Pinus* spp.), and spruces (*Picea* spp.).

    CONTROL: Pick off any larvae that you can reach. Spread sheets under infested trees in late spring to catch larvae as they fall; check the sheets daily and drop collected larvae into a bucket of soapy water. Sprays of superior oil can help, but do not use on blue spruces, since the oil will remove the blue color for that season.

■ Cucumber beetles

DESCRIPTION: Adult cucumber beetles are about ¼ inch long with greenish yellow, elongated bodies, black heads, and black stripes or dots on their wings. Larvae are slender white grubs.

DAMAGE: Adults chew large ragged holes in leaves, stems, and fruit. They also spread diseases, such as bacterial wilt and viruses, as they feed. Cucumber beetles are a serious pest on cucumbers, melons, and squash; they may also attack other vegetable and ornamental crops. Larvae feed on roots of corn and other crops, causing wilting and death.

CONTROL: Handpick adult beetles and drop them into a container of soapy water. Spray serious infestations with pyrethrins or rotenone. To minimize damage in following years, cover seedlings or transplants with floating row covers right after planting. Remove the cover as soon as flowers appear to allow for pollination. Apply parasitic nematodes to the soil to control larvae.

Cucumber beetle (actual size)

■ Cutworms

DESCRIPTION: Cutworms are plump, smooth, brown, gray, or green 1-inch caterpillars. They curl up when disturbed. Adults are brown or grayish moths.

DAMAGE: Cutworms curl around the base of young stems, chewing through vegetable and flower seedlings and transplants near the soil line. The larvae are only active at night and are most troublesome in spring.

CONTROL: Prevent damage by surrounding stems with cardboard collars extending 2 inches above and below the soil line. Handpick larvae. Apply parasitic nematodes to the soil around the base of young plants.

Cutworm (actual size)

▪ Fall webworms

DESCRIPTION: Larvae are slightly over an inch long, with a light brown, black-dotted body covered with yellow and white hairs. Adults are white moths.

DAMAGE: Starting in summer, fall webworms spin webs near branch tips and feed on the leaves within. Fall webworms damage foliage and produce their unsightly nests on a wide variety of deciduous trees and shrubs.

CONTROL: Prune off and destroy nests as they appear. Spraying leaves with BTK *(Bacillus thuringiensis* var. *kurstaki)* in late spring to early summer may help control larvae before they are protected by their webs.

Fall webworm (actual size)

▪ Flea beetles

DESCRIPTION: Adults are tiny, dark beetles that jump quickly when disturbed. The larvae are small white grubs that live in the soil.

DAMAGE: Adult beetles feed on leaves, leaving many tiny holes. Damage is most serious early in the season. Larvae feed on roots. Flea beetles attack many vegetable crops, including beans, broccoli, cabbage, corn, eggplants, peppers, potatoes, radishes, spinach, and Swiss chard.

CONTROL: Spray seriously infested plants with

Flea beetle (actual size)

neem, pyrethrins, or rotenone. Apply parasitic nematodes to the soil to control larvae. Prevent damage to future crops by protecting new plantings with floating rows cover until midsummer.

Fruit fly larvae and damage

- **Fruit flies**

   Description: Adult fruit flies look much like small houseflies with dark bars on their wings. The larvae are ¼-inch, creamy white, legless maggots.

   Damage: Adult fruit flies lay eggs under the skin of developing fruit, starting in midsummer. The eggs hatch into larvae, which tunnel through the flesh as they feed. Infested fruits may be small and deformed, and they often drop early. Cherry fruit flies attack cherries, pears, and plums. Related species, including currant fruit flies, Mediterranean fruit flies, and walnut husk flies, feed on blueberries, citrus, currants, and walnuts.

   Control: Once maggots are in the fruit, there is no control. The best strategy is to pick up and destroy dropped fruit daily during the summer to reduce damage to future crops. If damage was severe in past years, hang spherical red sticky traps in susceptible trees during June. As soon as you see adults on the traps, spray trees with rotenone.

- **Garden webworms**

    DESCRIPTION: These light green 1-inch caterpillars have dark spots along their sides. Adults are white to brown moths.

    DAMAGE: Webworms spin fine webbing over leaves and feed on the enclosed foliage. Severe infestations can weaken or kill plants. Many vegetable crops are susceptible; strawberries may also be attacked.

    CONTROL: Handpick caterpillars and drop them in a container of soapy water. Cut off and destroy webbed leaves or branches. Grow pollen- and nectar-producing plants to attract predators. Spray with BTK *(Bacillus thuringiensis* var. *kurstaki)* when caterpillars are still small. Use pyrethrins or rotenone as a last resort.

Garden webworm (actual size)

■ Grubs

DESCRIPTION: Grubs are plump, whitish, wormlike creatures, usually between 1/2 inch and 1 1/2 inches long. They often have dark heads. Grubs are the larvae of various beetles, including Japanese beetles and June beetles.

DAMAGE: Grubs feed on roots, weakening plants and causing wilting or death. Severe infestations can kill large patches of grass. Damage is usually most serious during fall and spring. Skunks, birds, and moles digging in your yard often indicate grub infestations. Grubs usually feed on lawn grasses but may also damage garden crops.

CONTROL: Milky disease is a biological control that attacks grubs. Use a fertilizer spreader to apply the spores evenly over your lawn. It may take several seasons to produce a noticeable effect, but once the spores build up, they will provide long-term control. For faster results, apply parasitic nematodes to affected areas.

Grub
(actual size)

- **Gypsy moths**

    Description: These caterpillars grow up to 2½ inches long, with a hairy gray or brown body marked with blue and red spots. Adult males are brown moths. Adult females are nonflying whitish moths that lay fluffy tan egg masses in late summer and fall.

    Damage: Young larvae feed on the edges of leaves starting in spring. As they grow, they begin to chew large holes in leaves. Severe infestations can defoliate trees. Deciduous trees may be able to produce another set of leaves that year, but repeated defoliation severely weakens growth. Evergreens may be killed. Gypsy moths attack many deciduous and evergreen trees; favorite targets include oaks (*Quercus* spp.), willows (*Salix* spp.), and lindens (*Tilia* spp.).

    Control: Handpick larvae that you can reach (wear gloves) and drop them in a container of soapy water. Spray leaves weekly with neem or BTK *(Bacillus thuringiensis* var. *kurstaki)* in mid- to late spring. In fall and winter, search for the egg masses on trunks, branches, or rocks; scrape them into a container of soapy water. Pheromone traps may help prevent mating and reduce future populations.

Gypsy moth
larva
(actual size)

■ Lace bugs

DESCRIPTION: Lace bugs are small, flat, square insects with lacy, transparent wings. Larvae are tiny and wingless, with spines.

DAMAGE: Adults and larvae feed on the undersides of leaves through much of the growing season, sucking plant juices and producing a yellow or whitish stippling. Infested leaves turn lighter in color, then may curl, turn brown, and drop. Lace bugs are a common problem on azaleas grown in full sun. These pests also attack many other vegetables, flowers, and shrubs.

CONTROL: Spray leaves thoroughly with insecticidal soap or neem. Treat serious infestations with pyrethrins or rotenone.

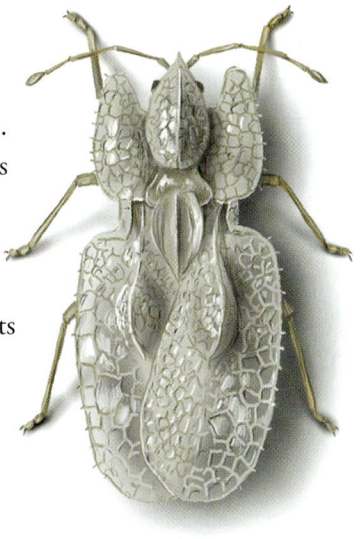

Lace bug
(actual size)

■ Leafhoppers

DESCRIPTION: Adult leafhoppers are small, greenish, wedge-shaped, soft-bodied insects that hop quickly when disturbed. Nymphs look similar to the adults but lack wings.

DAMAGE: Adults and nymphs feed on stems and the undersides of leaves. They suck plant juices, causing discoloration and stunted or distorted growth. The tips and sides of affected leaves may turn yellow or brown and curl upward. As they feed, leafhoppers exude a sticky honeydew on leaves and fruit. These pests may also transmit plant diseases. They attack most vegetables and fruits and some ornamental plants.

CONTROL: Use a strong spray of water to wash nymphs off plants. Spray serious infestations with insecticidal soap or pyrethrins; use rotenone or sabadilla as a last resort. In future years, grow resistant cultivars, or protect crops with floating row covers from planting time to flowering or harvest.

Leafhopper
(actual size)

Leafminer damage

▪ Leafminers

Description: Leafminers are small, pale, wormlike creatures found within leaves. Adults may be tiny black or black-and-yellow flies or black thick-bodied wasps.

Damage: Larvae feed between the upper and lower leaf surfaces through the growing season, producing papery blotches or meandering tunnels. The damaged areas may be yellow or brown. Infested leaves may drop. Leafminers feed on many vegetables and ornamentals, including beets, cabbage, chrysanthemums, columbines (*Aquilegia* spp.), nasturtiums, spinach, and Swiss chard; alders (*Alnus* spp.), birches (*Betula* spp.), and elms (*Ulmus* spp.) are also affected.

Control: If you see just one or two damaged spots on a leaf, cut out the damage with scissors. Remove and destroy badly infested leaves. To prevent damage in the future, protect vegetable crops with floating row covers from planting to flowering or harvest.

Leafroller web

- **Leafrollers**

DESCRIPTION: These yellowish to pale green ½- to 1-inch caterpillars hatch in clusters and feed at night. They hide during the day by rolling leaf edges together with webbing. Adults are golden or reddish brown moths.

DAMAGE: Leafrollers feed on leaves, buds, and fruit enclosed in their webs. Serious infestations can defoliate plants. Webbed leaves and shoot tips are unsightly. Damage is usually most serious in spring and late summer. Leafrollers attack a wide range of garden plants; some common targets include apples, roses, spireas (*Spiraea* spp.), and strawberries.

CONTROL: Pick off and step on rolled leaves. Sprays of BTK *(Bacillus thuringiensis* var. *kurstaki)* may help before leafrollers are protected by their webs.

■ Mealybugs

DESCRIPTION: Adult female mealybugs are small, white, oval, soft-bodied insects. Males are tiny winged insects. Nymphs look like the adult females but are even smaller.

DAMAGE: Adults and nymphs suck sap from leaves and shoots, causing off-color and stunted plants. As they feed, they exude sugary honeydew, which produces a sticky, shiny coating on leaves and supports the growth of dark sooty mold. Mealybugs attack many vegetables, ornamentals, fruit crops, and houseplants. They may be a problem any time during the growing season outdoors; indoors, they feed year-round.

CONTROL: Grow pollen- and nectar-rich plants to attract natural predators and parasites. On small plants, touch individual mealybugs with a swab dipped in isopropyl alcohol; discard severely infested plants. Spray larger plants with insecticidal soap or pyrethrins.

■ Mexican bean beetles

DESCRIPTION: Adults are copper-colored, round-backed, $1/4$-inch beetles with 16 black spots on their wing covers. They lay clusters of oval yellow eggs on the undersides of leaves. The eggs hatch into spiny, plump yellow grubs.

DAMAGE: Adults and larvae feed on the undersides of leaves, producing many holes and giving the leaves a lacy look. Severe infestations may defoliate plants. Damage is most serious in mid- to late summer.

CONTROL: Handpick adults and larvae and drop them in a container of soapy water; crush egg masses with your fingers. Spray serious infestations with neem, pyrethrins, or rotenone. To minimize damage to future crops, clean up garden debris in fall to remove overwintering sites. Look for resistant cultivars. Protect plants with floating row covers until the crop is established. Interplant beans with flowers and herbs to attract natural predators.

■ Mites

DESCRIPTION: Adult mites are very tiny golden, red, or brown spiderlike pests. Nymphs look much like adults but are even smaller and lighter in color. Some mites spin fine webs around leaves or between leaves and stems.

DAMAGE: Mites suck plant juices from leaves, producing a light-colored stippling on leaf surfaces. Whole leaves become pale and dry and may drop. Seriously infested plants may have webbing and stunted growth. Outdoors, mites feed through the growing season; indoors, they feed all year. Mites damage many vegetables, ornamentals, fruits, and houseplants.

CONTROL: Rinsing or spraying leaves frequently with water can suppress mite populations. Pollen- and nectar-rich plants attract natural predators, such as lady beetles and lacewings. Spray serious infestations with insecticidal soap, superior oil, neem, or pyrethrins.

Mite (actual size)

Mole cricket
(actual size)

- **Mole crickets**

    DESCRIPTION: Adults are large brownish insects that are 1 to 1½ inches long. They are covered with many short hairs and have flattened front feet that are well adapted to digging in the soil. Nymphs are similar but smaller.

    DAMAGE: Mole crickets tunnel under the ground and feed on roots. They usually damage lawns, causing irregular wilted, brown, or dead patches; they may also feed on the roots of garden plants. Mole crickets are a problem in southern gardens. Damage is most serious during warm, moist weather.

    CONTROL: Apply a solution of parasitic nematodes to the soil in infested areas, then water thoroughly.

- **Oriental fruit moths**

    DESCRIPTION: The larvae of oriental fruit moths are pinkish to grayish white ½-inch caterpillars with dark heads. Adults are gray-brown moths with brown markings on the wings.

    DAMAGE: The first larvae of the season tunnel into tender, rapidly growing shoots in early summer, causing shoot tips to wilt, turn brown, and die. Larvae that hatch later in the season enter the fruit and feed on the flesh, usually near the center. Oriental fruit moths attack many fruit trees and some ornamental trees.

Oriental fruit moth
larva and damage

CONTROL: Pick and destroy dropped fruit. Use pheromone traps to monitor and control adults. Spray superior oil to kill the eggs and larvae. Cultivate the top 3 inches of soil just before trees bloom in spring to expose overwintering larvae. Grow flowering groundcovers under fruit trees to attract natural predators. If oriental fruit moths are a common pest in your area, prevent damage by planting early-ripening cultivars.

Pear sawfly larvae and damage

■ Pear sawflies

DESCRIPTION: The larvae of pear sawflies are commonly known as pear slugs, since they look much like ½-inch, greenish brown, large-headed slugs. Adult pear sawflies are small black and yellow flies that emerge in late spring.

DAMAGE: Pear slugs feed on the upper sides of leaves, rasping off the leaf surface or chewing completely through leaf tissue between the veins. Affected leaves may have scorched-looking patches or a lacy appearance. Most serious damage occurs in early and late summer. Pear sawflies attack many rose-family plants, including cherries, cotoneasters (*Cotoneaster* spp.), pears, and plums.

CONTROL: Handpick the slimy larvae that are within reach (wear gloves if you're squeamish!) and drop them in a container of soapy water. Spray serious infestations with insecticidal soap, pyrethrins, or rotenone. Cultivate around trees in spring to expose overwintering larvae to predators.

Tarnished plant bug (actual size)

- **Plant bugs**

    DESCRIPTION: Adult plant bugs are fast-moving, oblong, flattened insects, $1/4$ to $1/3$ inch long. Four-lined plant bugs are greenish yellow, with four black stripes on the back. The wingless nymphs are reddish with black dots. Tarnished plant bugs are greenish to brownish, with brown or black mottling on the back. Nymphs are smaller and pale yellow with black dots; they lack wings.

    DAMAGE: Plant bugs suck plant juices, causing sunken brown or black spots on leaves and deformed leaves, buds, and shoots. They may also feed on fruit, producing scarring, swellings or depressions, and distorted growth. Plant bugs are active through spring and summer. They attack many garden plants; asters, dahlias, mint, peaches, strawberries, and Swiss chard are just a few common hosts.

    CONTROL: Handpick adults and nymphs in early morning (while they are still sluggish) and drop them in a container of soapy water. Grow pollen- and nectar-rich plants to attract natural predators. Treat serious infestations with neem, sabadilla, or rotenone. To prevent damage to future crops, clean up garden debris in fall and spring to remove overwintering sites for adults. Floating row covers are a good way to protect young vegetable crops from plant bugs.

Plum curculio
(actual size)

- **Plum curculios**

    DESCRIPTION: Adults are hard-shelled, brownish gray, ¼-inch beetles with a downward-curving snout. The larvae are plump, white, brown-headed grubs.

    DAMAGE: Adults feed on leaves and blossoms and lay eggs in fruit through summer, producing crescent-shaped scars in the skin. The larvae tunnel in the flesh of the fruit. Infested fruit often drops early. Plum curculios attack many fruits, including apples, blueberries, cherries, peaches, pears, and plums.

    CONTROL: Starting at blossom time, spread sheets under infested trees in early morning and tap the branches with a padded stick; collect the curculios that fall and drop them in a container of soapy water. Repeat daily for four to six weeks to control adults. Pick up and destroy dropped fruit every day to prevent larvae from entering the ground to pupate.

Scale insects
(actual size)

- **Scale insects**

    DESCRIPTION: Adult female scales are small, waxy, soft- or hard-bodied, stationary insects. They may be red, white, brown, black or gray. Adult males are tiny and winged. Nymphs are smaller than adults and have legs; they can move around for a short time before molting to the adult form.

    DAMAGE: Scale insects suck plant juices from shoots and leaves, causing stunted, off-color growth. Severe infestations may cover a large portion of the stem. As they feed, scale insects exude a sticky honeydew that supports the growth of dark sooty mold. These pests attack many garden and indoor plants. Outside, they may be present through most of the season; indoors, they feed all year.

CONTROL: Pinch or prune off and destroy scale-infested growth. Use a fingernail or a soft brush and soapy water (rinse plants afterward) to remove scale insects from stems. Wipe leaves with a wet cloth to remove the honeydew and sooty mold. Plant pollen- and nectar-rich plants to attract natural predators. Spray infested trees with superior oil. Treat serious infestations with pyrethrins or rotenone as a last resort.

Slug
(actual size)

■ Slugs and snails

DESCRIPTION: Adults are gray, tan, or black, slimy, soft-bodied mollusks. Snails have a hard outer shell and may be up to 1½ inches long. Slugs lack shells; they may be ⅛ inch to 6 inches or more in length. Both snails and slugs leave slime trails on leaves. Slugs are usually most active at night and in damp places; snails are less dependent on moisture.

DAMAGE: Slugs and snails rasp large holes in leaves, stems, and fruit; they may completely devour seedlings. These pests are active throughout the growing season and will attack almost any plant.

CONTROL: Trap slugs and snails under fruit rinds, cabbage leaves, or boards set on the soil, or in shallow pans of beer set into the soil surface; check traps daily and destroy pests. Eliminate mulches and garden debris; these materials provide ideal hiding places. Use barriers of copper screen or sheeting to repel slugs and snails. Plant ground covers to attract ground beetles and other predators.

Sod webworm (actual size)

▪ Sod webworms

DESCRIPTION: Sod webworms are ½- to 1-inch long, gray, cream, or light brown caterpillars with dark-spotted backs. Adults are whitish or gray moths that fly in a zigzag pattern.

DAMAGE: Larvae feed on leaf blades of grass plants, producing small, irregular brown or dead patches. They also produce silken tunnels in the ground. Damage is most serious in mid- to late summer, especially during hot, dry conditions. Sod webworms attack most turf grasses but are particularly attracted to bentgrasses and bluegrasses.

CONTROL: Drenching infested areas with BTK *(Bacillus thuringiensis* var. *kurstaki)* or parasitic nematodes may help control these pests. If sod webworms are a problem every year, consider overseeding or reseeding with endophyte-containing turfgrass cultivars.

Spruce budworm (actual size)

- **Spruce budworms**

    DESCRIPTION: Larvae are dark brown, cream-spotted caterpillars up to ¾ inch long. Adults are gray-spotted brown moths.

    DAMAGE: Larvae feed on buds and needles from spring to early summer, often webbing several needles together and chewing them off at the base. Infested shoots may wilt. Damaged trees are unsightly and can be seriously weakened. Spruce budworms are serious pests on several conifers, including firs (*Abies* spp.), hemlocks (*Tsuga* spp.), pines (*Pinus* spp.), and spruces (*Picea* spp.).

    CONTROL: Cut off and destroy infested tips. If the center shoot (leader) must be removed, create a new leader by training a lower, undamaged shoot up a stake fastened to the main stem. Use pheromone traps to control adult moths. Spray serious infestations with BTK *(Bacillus thuringiensis* var. *kurstaki)*.

Stink bug
(actual size)

■ Stink bugs

DESCRIPTION: Adults are ½-inch long, green, gray, or brown, triangular-shaped bugs. Nymphs are similar but wingless. The harlequin bug is a type of stink bug. It is black with red or yellow markings on the back.

DAMAGE: Stink bugs suck plant juices from leaves, stems, buds, and fruit. Damaged foliage may be yellowed or brown; injured fruits are scarred and deformed. These pests feed through the summer in most areas; they may be active all year in mild-winter areas. Stink bugs attack many garden plants, including beans, okra, peaches, peas, and tomatoes. Harlequin bugs are especially fond of cabbage, turnips, and related crops.

CONTROL: Handpick adults and drop them in a container of soapy water. Grow pollen- and nectar-rich plants to attract natural predators. Treat serious infestations with pyrethrins or sabadilla. Clean up garden debris to remove overwintering sites.

■ Tent caterpillars

DESCRIPTION: Larvae are dark caterpillars with a white stripe on the back and blue or reddish marks on the sides. They are hairy and may reach 2 inches long when mature. These caterpillars form weblike tents or bags in crotches of tree branches. Adults are reddish brown moths. Adult females lay bands of foamy-looking gray-brown egg masses on twigs in midsummer.

DAMAGE: Larvae are active in late spring to early summer, consuming leaves on affected branches. In extreme cases they may defoliate the tree. Tent caterpillars feed on many

Tent caterpillar
(actual size)

deciduous trees and shrubs, especially apples, black cherries, pears, and plums.

CONTROL: Tear down nests with a stick or prune them out; burn or dispose of the webs. Prune off and destroy egg masses. Sprays of BTK *(Bacillus thuringiensis* var. *kurstaki)* may be effective in early to midspring, while caterpillars are still small.

▪ Thrips

DESCRIPTION: Adults are tiny, slender, brown, yellow, or black insects with narrow fringed wings. Nymphs are similar but wingless and even smaller.

DAMAGE: Adults and nymphs feed on leaves, flowers, buds, and stems, causing browning and white flecking and deforming flowers, buds, and leaves. Thrips attack a variety of indoor and outdoor plants, including beans, gladiolus, onions, and squash. Outside, they feed through the growing season; indoors, they may feed all year.

CONTROL: Remove and destroy infested flowers and buds. Spray with insecticidal soap or superior oil, or dust with diatomaceous earth in the evening; treat serious infestations with pyrethrins.

Thrip damage

Black
vine weevil
(actual size)

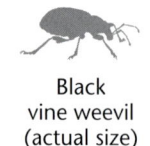

■ Weevils

DESCRIPTION: Adult weevils are small hard-shelled beetles with long snouts; they are usually 1/8 to 1/4 inch long and tend to be most active at night. Larvae are plump, whitish, dark-headed, legless grubs. Many weevils are garden pests; a few of the most common include black vine weevils, strawberry weevils, and vegetable weevils. Adult black vine weevils are brownish black. Strawberry weevils are reddish brown. Vegetable weevils are light brown.

DAMAGE: Adult weevils chew notches in leaves, stems, and buds in summer; larvae tunnel and feed in roots or seeds in fall and spring. Weevils attack many edible and ornamental plants. Black vine weevils are serious pests of rhododendrons and yews (*Taxus* spp.). Strawberry weevils chew on the crowns, leaves, and flowers of strawberries. Vegetable weevils attack cabbages, carrots, radishes, Swiss chard, and other vegetable crops.

CONTROL: Drench the soil around plants with parasitic nematodes to control larvae. Handpick adults, or treat serious infestations by dusting plants with pyrethrins or rotenone in the evening.

Whiteflies
(actual size)

■ Whiteflies

DESCRIPTION: Adults are tiny flies with white powdery wings. They cluster on the undersides of leaves and fly up in great numbers when disturbed. Larvae are tiny and flattened.

DAMAGE: Adults and nymphs suck plant juices through the season outdoors and all year indoors. Infested plants look yellow, sickly, and stunted. As they feed, whiteflies exude a sticky honeydew that supports the growth of black sooty mold. Whiteflies feed on many vegetables and ornamentals and are especially common in greenhouses.

CONTROL: Indoors, catch whiteflies on sticky yellow cards or use a handheld vacuum to suck pests off plants. Spray serious infestations with insecticidal soap, superior oil, pyrethrins, or rotenone.

■ Wireworms

DESCRIPTION: Wireworms are slender, shiny, orange grubs up to 1½ inches long. They are the larvae of dark, elongated, 1½-inch adults known as click beetles.

DAMAGE: Wireworms chew on crowns, corms, roots, and seeds, causing wilting or stunted growth. Wireworms are most serious in gardens recently made from grassy areas. Most damage occurs in spring and fall. These pests feed on many vegetable crops, as well as gladiolus corms.

CONTROL: Make wireworm traps from pieces of potato stuck onto pointed sticks. Bury the potato pieces in the soil in early spring; use the sticks to lift them every day or two and remove and destroy trapped larvae. Apply parasitic nematodes to the soil. If possible, delay spring planting until the soil is warm.

Wireworm (actual size)

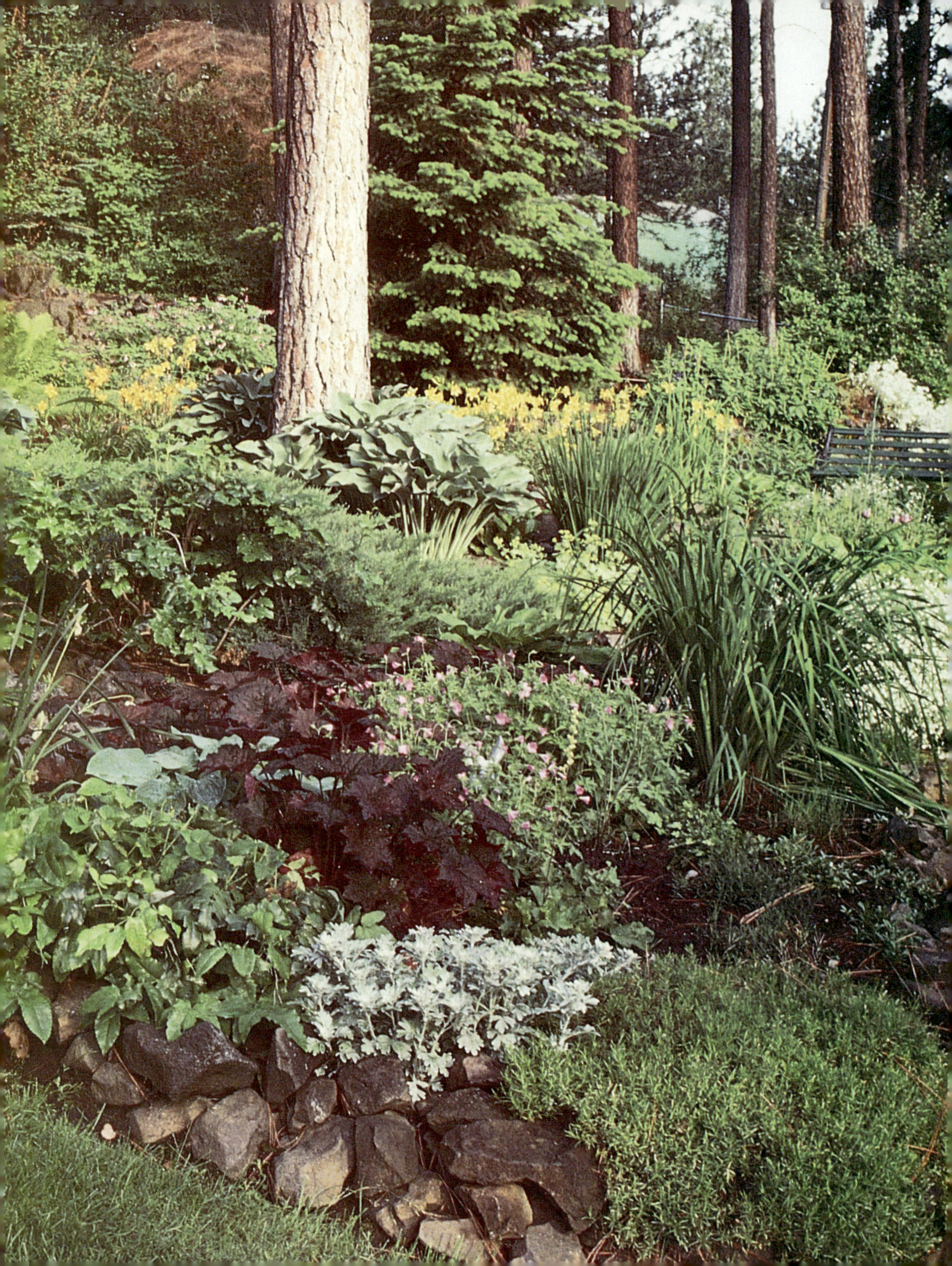

Chapter 3:
A Guide to Garden Diseases

Diseases can be transmitted to your plants quite rapidly by a variety of means — air, soil, water, insects, and animals. Gardeners can also carry them from plant to plant on their clothing and tools. Beacuse pathogens are rarely visible, they are hard to control by mechanical or physical means. As a result, prevention plays a very important role in controlling plant diseases: it's much easier to protect a plant from disease than to cure it. Fortunately, preventing disease problems is fairly straightforward. Building healthy soil, planting resistant cultivars, and rotating crops are three techniques that are effective. Review the steps in Chapter 1 for a complete rundown on preventing diseases by creating optimum conditions for your plants. If you do notice a disease developing in your garden, refer to the following pages to identify the problem and learn how to handle it.

Preventing diseases can be as pretty as it is practical. These lush perennial beds keep lawnmowers away from the tree trunks, thus preventing damage that can lead to disease and pest problems.

■ Apple scab

Description: Symptoms of this fungal disease first appear as pale yellow spots on the undersides of leaves. These spots gradually darken and expand, spreading to upper leaf surfaces and blossoms. Affected leaves may crack or curl and drop. Spots also appear on fruit and develop into dark scabs; fruit may be cracked and deformed.

Plants Affected: Apples and crabapples.

Control: Clean up and destroy all diseased and dropped fruit and leaves. Prune to keep the center of the tree open to light and air. Spray weekly with lime-sulfur from the first sign of green in the buds until blossoms start to open if weather is warm (above 60°F) and damp. Look for scab-resistant or tolerant cultivars.

■ Black spot

Description: This fungal disease appears as black leaf spots with feathery margins. Leaves turn yellow between spots and drop early. Dark cankers may appear on stems.

Plants Affected: Roses.

Control: Pick off and destroy infected leaves; prune out affected canes. At the first sign of the disease, spray thoroughly every seven to ten days with fungicidal soap or sulfur. Or use a baking soda spray, made from 1 teaspoon of baking soda, 1 quart of water, and a few drops of liquid dish soap. To minimize problems in future, prune plants to improve air circulation and avoid overhead watering. Plant resistant cultivars.

■ Blossom end rot

Description: Damage starts with a water-soaked spot or bruise on the end of the fruit. The spot gradually turns a tan color, then becomes dark, flattened, and leathery.

Plants Affected: Tomatoes and peppers.

Control: Blossom end rot is a cultural problem caused by a calcium deficiency. It is often related to temperature extremes, uneven watering, or root damage. Minimize further damage by mulching to keep soil evenly moist and eliminate the need to cultivate around roots. Spraying with seaweed extract may provide a quick boost of calcium to affected plants. Take a soil test in fall and add lime, if needed, to correct the deficiency.

- **Brown rot**

    DESCRIPTION: During warm, wet spring weather, this fungal disease infects blossoms, causing them to turn brown, wilt, and drop. Sunken cankers form on twigs near branch tips, and branch tips may die. Later in the season, fruit develops small round brown spots that quickly spread to cover the whole fruit. Rings of grayish spores may be visible in browned areas. Spots spread until the whole fruit rots and shrivels and hangs on the tree.

    PLANTS AFFECTED: Many fruit trees, including apricots, cherries, peaches, and plums; sometimes apples and pears.

    CONTROL: Pick off and destroy shriveled fruit. Prune away damaged and dead twigs. Apply a preventive sulfur spray before bloom and again seven to ten days after bloom. Prune to open the center of the tree to light and air. Plant resistant cultivars.

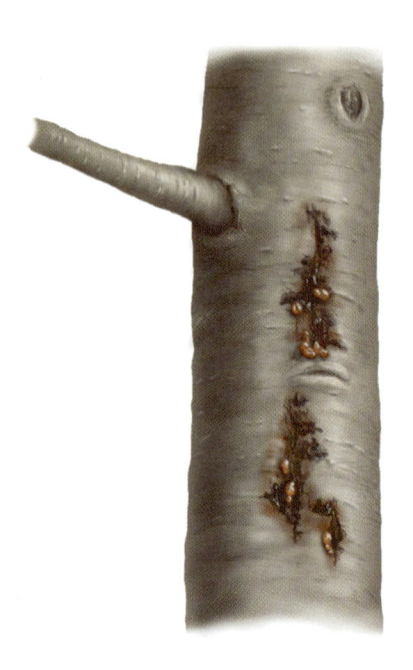

- **Canker**

    DESCRIPTION: Fungal and bacterial cankers produce discolored spots and dead areas on stems. Affected areas may be covered with small black spores or ooze a slimy or gummy material; the slime or gum may smell sour if bacteria have caused the canker. Shoot tips may turn yellow, wilt, and die.

    PLANTS AFFECTED: Many trees and shrubs, especially apples, maples (*Acer* spp.), peaches, pears, plums, roses, and willows (*Salix* spp.).

    CONTROL: During dry weather, cut off infected parts at least 2 inches below the affected area. Disinfect pruners after each cut by dipping them in a bleach solution (1 part bleach to 9 parts water). Remove and destroy severely diseased plants. To prevent future cankers, avoid wounding bark with pruning or lawn-care tools.

■ Cedar-apple rust

DESCRIPTION: On junipers, this fungal disease starts as small brown swellings on needles. The swellings gradually expand to become hard brown galls. During warm, wet spring weather, galls absorb water and produce long, orange, spore-producing "horns." When the spores infect apples or crabapples, they produce yellow to orange spots on upper surfaces of apple leaves and on fruit. These spots release spores that then infect junipers.

PLANTS AFFECTED: Eastern red cedar *(Juniperus virginiana)* and some other species of junipers; apples and crabapples.

CONTROL: Remove and destroy galls and infected foliage and fruit. Sprays are generally not of much use. Keeping apples and junipers at least 3 miles apart can be an effective, although usually not practical, way to prevent the disease. If you know cedar-apple rust is a problem in your area, plant rust-resistant species and cultivars.

■ Club root

DESCRIPTION: The first symptoms of this fungal disease include wilted and yellow leaves and stunted plants. Infected plants may yield poorly or die. The roots of these plants are gnarled and distorted, with thick galls.

PLANTS AFFECTED: Cabbage-family plants.

CONTROL: Remove and destroy infected plants. Keep soil pH near neutral (above 7.2 if possible). Improve drainage. When planning crop rotations, leave at least seven years between susceptible crops.

■ Crown gall

DESCRIPTION: This soil-borne bacterial disease produces rounded, knobby growths on stems, usually near the soil line. Galls may also appear on roots and branches. Affected plants grow slowly and may die. Fruits are small and may be deformed.

PLANTS AFFECTED: A wide range of garden plants, especially apples, brambles, euonymus, forsythia, pears, and roses.

CONTROL: Try pruning off the galls, disinfecting your shears between cuts with a bleach solution (1 part bleach to 9 parts water). If galls return, remove and destroy infected plants. Improve soil drainage. Avoid replanting with susceptible crops. Inspect plants for signs of the disease before buying them.

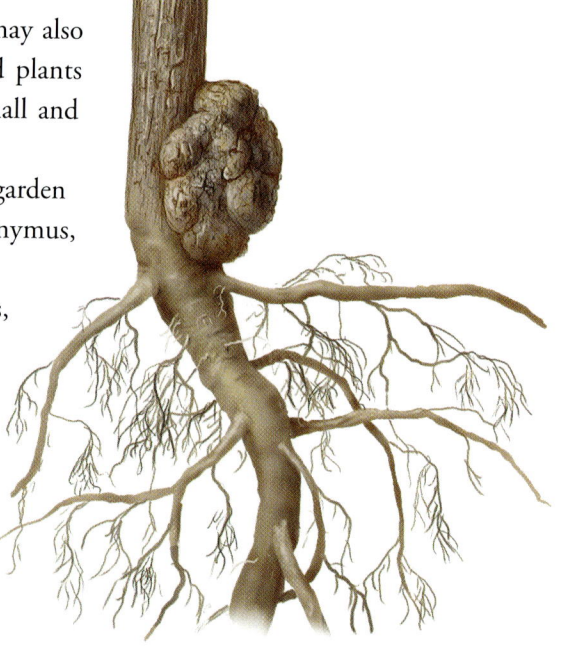

■ Damping-off

DESCRIPTION: Damping-off fungi may attack seeds even before they sprout, causing them to turn brown and mushy. Infected seedlings rot at the soil line and fall over.

PLANTS AFFECTED: Seeds and seedlings.

CONTROL: Moving pots or trays to a warm, dry place may stop the spread of the disease. However, the best control is prevention. Start seed in a disease-free medium, such as vermiculite or sterilized seed-starting mix. Sow thinly to avoid overcrowding. Sprinkle a light layer of milled sphagnum moss over sown seeds. Provide seedlings with good air circulation. Water pots from below (by setting them in trays of water until the mix is moist) to avoid wetting the surface.

■ Fairy ring

Description: Fairy ring fungi produce circular or nearly circular areas of grass that grow faster and are greener than surrounding areas. Nearby turf may be dead or grow poorly. Mushrooms or puffballs — the spore-producing bodies of the fungus — may appear in the rings. Fairy rings expand by several inches to several feet per year.

Fairy rings appear in lawn grass that is growing over a rotting tree stump. Removing the stump is the best, most permanent, solution. Another option would be to remove the grass and plant the area with mixed perennials and shrubs.

PLANTS AFFECTED: Turf grasses.

CONTROL: If you know of any wood pieces or tree stumps buried in the area, dig them up and remove them. Break or mow off the fruiting bodies. Fertilizing and watering may stimulate surrounding grass, producing a more even green color to mask the rings. Top-dress the lawn with compost to encourage beneficial soil organisms. Aerate the soil in and around the ring with a spading fork; wash the tool thoroughly before using it elsewhere.

■ Fire blight

DESCRIPTION: This bacterial disease causes flowers to wilt and die. Stem tips turn black and wilt, developing a distinctive curled shape. Leaves turn brown, die, and cling to branches. Stems develop oozing reddish brown stem cankers that later turn brown and dry. Fire blight can spread quickly in warm, wet weather to kill entire plants.

PLANTS AFFECTED: Many rose-family plants, including apples, firethorns (*Pyracantha* spp.), and pears.

CONTROL: Prune out and destroy diseased shoots as soon as you see them, cutting at least 6 inches below the visibly damaged area. Disinfect shears in a bleach solution (1 part bleach to 9 parts water) between cuts. Cut down and destroy severely infected plants in winter. To prevent problems, spray with bordeaux mix when the tree is dormant or with streptomycin just prior to bloom and when plants are in full bloom. Plant resistant cultivars.

Gray mold

Description: This fungal disease, also known as botrytis blight, produces gray-brown spots or streaks on leaves, buds, and flowers. A woolly fungal growth may be visible in the spots. Affected parts may turn brown and wither. Fruits develop a fuzzy gray mold, then rot.

Plants Affected: Many edible and ornamental plants.

Control: Remove and destroy infected plant parts. Thin crowded stems and space new plants properly for good air circulation. Pick often to avoid overripe fruit. Handle fruit gently to prevent bruising.

Leaf scorch

Description: Leaf edges become yellow, then turn brown and crispy; they may also roll inward. Leaves or whole plants may wilt. Affected plants have stunted growth.

Plants Affected: Any plants growing in dry, hot sites, especially those exposed to heat reflected from walls and paving.

Control: This cultural problem is due to drought and excessive heat. To minimize damage, water susceptible plants regularly to maintain soil moisture. Mulches will also help to keep the soil moist and cool. When planting hot, dry sites, look for plants that are naturally adapted to those conditions. Work plenty of organic matter into the soil before planting.

■ Leaf spots

Description: Fungal and bacterial leaf spots can vary widely in appearance. Anthracnose produces dark, sunken lesions, often with pinkish spore masses. Bacterial spot leads to small, circular, light green spots that later turn brown; there may be angular purplish areas around the spots. Foliage infected by cherry leaf spot shows small, reddish or purple, circular spots with centers that drop out. Early blight produces small circular yellow spots that expand and turn brown with a grayish center; late blight forms fast-spreading purplish or brown-black areas, sometimes with pale halos. These and other leaf spot diseases may also produce sunken lesions on stems, stunted growth, and spotted or deformed fruit.

Plants Affected: A wide range of edible and ornamental plants.

Control: Pinch off and destroy infected parts. Clean up dropped leaves. Applying a sulfur- or copper-based fungicide every seven to ten days may prevent the spread of fungal leaf spots. Destroy severely infected plants. Prevent problems by growing resistant cultivars and by spacing and pruning for good air circulation.

▪ Mildew

DESCRIPTION: Powdery mildew and downy mildew are fungal diseases that produce white to gray patches on plant leaves, stems, buds, and flowers. Downy mildew also produces light green patches on upper leaf surfaces or cottony purplish lesions on leaves and stems. Mildew-infected leaves may be distorted and drop early. Fruits may be dwarfed and russeted or distorted.

PLANTS AFFECTED: Many edible and ornamental plants, especially bee balm *(Monarda didyma),* cucumbers, grapes, lettuce, lilacs (*Syringa* spp.), melons, phlox, roses, spinach, squash, and Swiss chard.

CONTROL: Remove infected leaves. Thin crowded stems to improve air circulation. Spray with a sulfur- or copper-based fungicide every seven to ten days. Destroy seriously infected plants. To prevent damage, grow resistant cultivars, space plants properly, avoid overhead watering, and avoid handling wet plants.

▪ Nitrogen deficiency

DESCRIPTION: Yellow leaves are a common symptom of nitrogen deficiency. The oldest leaves are the first to turn yellow; eventually all the leaves and buds turn off-color. Flowers, buds, and leaves drop early. Deficient plants grow slowly and may have low yields.

PLANTS AFFECTED: Any plant, but most often vegetable crops.

CONTROL: For a quick fix, spray foliage with fish emulsion, and add a high-nitrogen fertilizer, such as blood meal, to the soil. For future crops, work nitrogen fertilizers and compost into the soil before planting.

■ Peach leaf curl

DESCRIPTION: This fungal disease causes new leaves to be blistered, thickened, and distorted. Affected foliage may curl up and turn reddish. Blossoms and young fruits may also be damaged and drop early. Twigs may be stunted and swollen.

PLANTS AFFECTED: Nectarines and peaches.

CONTROL: Remove and destroy infected leaves. In fall, spray with lime-sulfur or bordeaux mix after leaf drop; spray again just before buds open in spring. Plant resistant cultivars.

■ Rots

DESCRIPTION: Fungal and bacterial rots cause wilted, off-color plants. Roots of affected plants are dark and dry or mushy, rather than firm and white.

PLANTS AFFECTED: Any plants, particularly those growing in water-logged soil.

CONTROL: Remove and destroy affected plants. Improve soil drainage. Look for resistant cultivars. Wait until the soil is warm to set out transplants. Set plants so the crown is slightly higher than the surrounding soil. Mulch to prevent root damage from cultivation.

■ Rust

Description: This fungal disease produces whitish or yellowish spots on the tops of leaves. Powdery yellow to orange spots appear on the undersides of leaves or on stems. Infected leaves turn yellow, dry up, and drop early.

Plants Affected: A wide range of edible and ornamental plants, including asparagus, beans, brambles, roses, and snapdragons *(Antirrhinum majus)*.

Control: Remove and destroy infected leaves. Dust with sulfur every seven to ten days until the disease is controlled. Clean up and discard plant debris in fall. To prevent problems, grow resistant cultivars. Mulch plants and use drip irrigation to keep leaves dry.

■ Viruses

Description: Viruses produce crinkled, deformed leaves, often with yellow-green mottling. Plants are stunted or unusually bushy and often do not set fruit; fruit that does form may be deformed or dry and flavorless.

Plants Affected: A wide range of edible and ornamental plants.

Control: There is no cure for viruses; remove and destroy infected plants. Rinse hands and tools with milk after handling virus-infected plants. Protect crops with row covers or use sprays to control aphids, cucumber beetles, leafhoppers, whiteflies, and other insects that spread viruses as they feed.

- **Wilts**

DESCRIPTION: Plants infected with fungal or bacterial wilt may have yellow or browned leaves; sometimes the leaves curl upward or drop early. Wilting and stunted growth are other common symptoms. Wilting may occur over the whole plant or just on one side. Bacterial wilts often produce a white, sticky material that oozes from cut stems. Dark lesions may form on stems at ground level, and roots may rot. Seriously infected plants may die.

PLANTS AFFECTED: A wide range of edible and ornamental plants.

CONTROL: Remove and destroy infected plants immediately. To prevent damage, grow resistant cultivars. Protect plants with row covers or use sprays to control cucumber beetles, flea beetles, and other insects that spread the disease as they feed. Use mulch around plants to keep the soil cool.

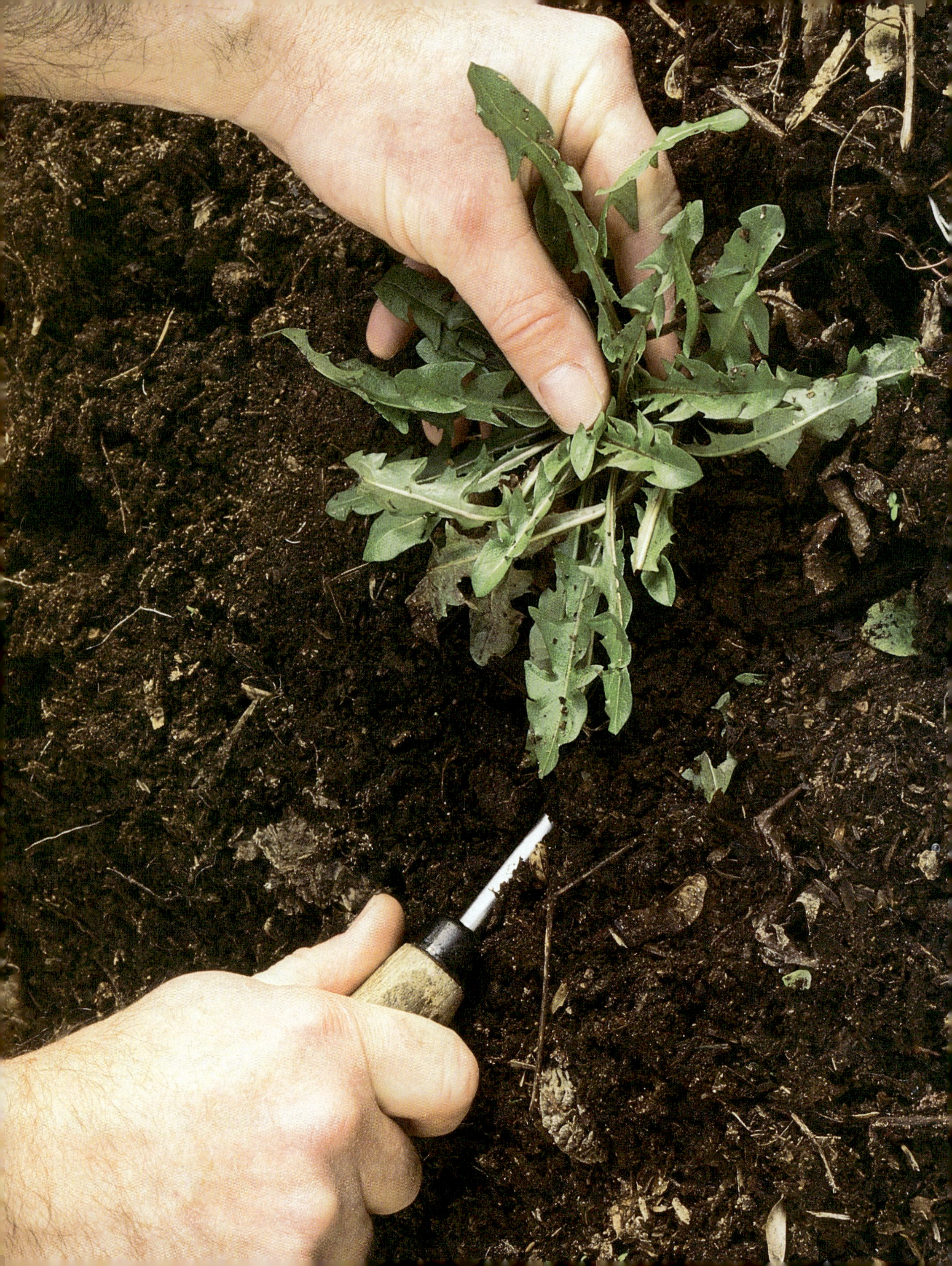

Chapter 4:
Dealing with Weeds

Weeds may be less damaging to your plants than insects or diseases, but keeping them under control is still important. Not only are weeds unsightly, they compete with garden plants for space, light, water, and nutrients. Weeds can also host insects and diseases that attack garden plants.

Fortunately, weed control doesn't have to be difficult. Using the preventive measures and safe, natural controls discussed in the following pages can make your weed woes a thing of the past. You'll also find a guide to 18 common lawn and garden weeds, with descriptions and specific control tips for each.

Preventing Weed Problems

The easiest way to handle weeds is to stop them before they get started. Prevention is an ongoing process, beginning from the time you prepare the soil for

---

To control perennial weeds most effectively, remove as much of the root as possible. Otherwise, weeds will resprout. Asparagus forks are useful for digging dandelions, which have deep taproots. Weeding is easiest when the soil is moist.

planting. Before digging or tilling, remove as many of the existing weeds as possible. Get as much of the root system as you can, especially from perennial weeds that spread by creeping roots or stems. Otherwise, digging or tilling will chop up and spread the creeping parts through the soil, leaving you with a bigger weed problem than before.

If the site was very weedy, you might want to take the extra step of solarizing it. To solarize, prepare the soil as you would for planting, then water it thoroughly and stretch a sheet of clear plastic over the site. Leave the plastic in place for six weeks if the weather has been warm and sunny, or up to ten weeks if the weather has been cool or cloudy. The heat that builds up under the plastic will kill many weed seeds and seedlings, as well as some insects and diseases. Once you remove the plastic, disturb the soil as little as possible during planting to avoid exposing buried weed seeds to the sun, which encourages germination.

When it isn't practical to wait so long before planting, you can try another technique. Prepare the soil for planting, water thoroughly, and wait for seven to ten days to let the surface weeds sprout. Hoe or till shallowly to eliminate those weeds. Repeating the process once or twice more before planting will be even more effective in discouraging weeds.

> **TIPS FOR SUCCESS**
>
> Repeated hoeing can reduce or eliminate weed problems in a new garden bed. Water the soil thoroughly to encourage weed seeds to germinate, wait for seven to ten days, and then hoe down the weeds. Repeat the process once or twice for even more effective control.

Your plants can also help control weeds for you. Space bushy crops so the plants just touch on each side, making a solid leafy cover that shades the soil and discourages weed seeds from sprouting. Or grow low, spreading plants as ground covers under more upright plants for the same effect.

Mulches are a major weapon in the war against weeds. They block sunlight from reaching the soil and smother weed seeds. They also help to keep the soil moist and loose, making it easier to pull out any weeds that do emerge. In the vegetable garden, looks aren't especially important, so you can use coarse mulches such as straw or chopped leaves. Flower beds, shrubs, and trees are more visible, so choose a finer material, such as shredded bark.

Weed the site thoroughly before mulching; mulch won't do much to smother existing weeds. A 2-inch layer works well with most mulches; coarse mulches can be 4 to 6 inches thick. Leave an inch or two of clear space around each plant stem to avoid rot.

DEALING WITH WEEDS

CONTROLLING GARDEN WEEDS

Sometimes weeds can sneak past even the best defenses. When they do appear, make every effort to get them while they are young. Hand-pulling a few seedling weeds each time you are out in the garden will save you hours of serious weeding later in the season. It's especially important to catch weeds before they bloom and set seed. If you don't have time to thoroughly weed a garden bed, a quick

Sometimes the best way to beat weeds is to join them. Try planting weed-prone areas with vigorous annuals or perennials. This walkway does have weeds in it, but Johnny jump-ups compete effectively with them and create an appealing effect overall.

solution is to pick off and discard weed flowers and seeds to prevent self-sowing. Leafy weeds can go in the compost pile. Discard or bury weeds that have flowers or seeds; otherwise, you'll just be spreading the weeds back on your garden when you spread the compost.

Physical controls. For easy weeding, wait until after a rain; it's easier to pull or dig most weeds out of moist soil. Try to get the roots, too, especially from weeds with thick taproots or creeping root systems. If you use a hoe or cultivator, try to stir up the soil as little as possible to avoid bringing more weed seeds to the surface. While these tools mostly leave the roots in the soil, repeated cultivation (every seven to fourteen days) can weaken and eventually kill even persistent weeds.

Another physical control for weeds is known as flaming. Flaming uses a propane torch that has been specially adapted for garden use. Suitable flaming tools are available through several garden supply catalogs.

Flaming works best on young, soft-stemmed weeds. It is especially useful for controlling weeds in paved areas. Don't use it during very dry weather or around flammable mulches, such as pine needles.

To use the tool effectively, sweep the flame over the tops of the weeds you want to control. Just a few seconds will do; the plants shouldn't look burnt. Within a few hours, treated weeds will begin to wilt and die. If weeds resprout from the roots, treat them again every week or two.

Natural sprays. When all other measures fail, you may resort to sprays for weed control. Commercial, organic, soap-based herbicides work best on young weeds; older weeds may need several treatments. Spray leaves thoroughly. These materials can kill any plant they touch, so use a piece of cardboard to prevent the spray from drifting onto desirable plants.

Other organic control options include hot water, vinegar, and salt. Because these materials can affect the soil as well as the weeds, they work best in paved areas, where you don't want any plants to grow. Simply dump boiling-hot water over the weeds (wear gloves to protect your hands), or pull off the tops and douse the roots with vinegar or salt.

DEALING WITH WEEDS

Lawn grass can become a troublesome weed if it wanders into a flowerbed. An edging strip like this one will make it stay put. Edging strips also make it easy to mow and trim in one easy step.

- ### Bindweeds

    DESCRIPTION: Bindweeds (*Convolvulus* spp.) are perennial weeds that reproduce by seed and spreading roots. The stems grow 6 feet long or more, and they may sprawl over the ground or wrap their twining stems around nearby plants to climb upward. The green leaves are usually triangular or heart-shaped. Pink or white funnel-shaped flowers bloom along the stems from summer into early fall.

    CONTROL: Pull seedlings as soon as you spot them; mulch bare soil to discourage more seeds from sprouting. To control established plants, dig them out, including as much of the root system as possible. Cut down or pull out new shoots that appear every week or two. Or cut plants to the ground, then mulch with heavy cardboard or a thick layer of newspaper; leave the mulch in place at least one growing season.

- **Broadleaf plantain**

    DESCRIPTION: Broadleaf plantain *(Plantago major)* is a perennial weed that reproduces by seed. It grows in low rosettes of oval green leaves with prominent veins on the undersides of the foliage. During summer and early fall, narrow, 4- to 12-inch spikes of small green flowers arise from the center of the rosettes.

    CONTROL: Pull plants or cut them off at soil level every week or two. Broadleaf plantain thrives in heavy, compacted soil, so aerate the site and add organic matter to loosen the soil; this will discourage the weed from returning.

■ Canada thistle

Description: Canada thistle *(Cirsium arvense)* is a perennial weed that reproduces by seed and by vigorously creeping roots. Its upright stems branch near the top and can grow up to 4 feet tall. The stems carry long lobed leaves with spiny edges. From summer into fall, the stems are topped with brushy, rosy-purple flowers. These flowers mature into heads of silky-tufted seeds that are carried by the wind.

Control: Pull seedlings (wear gloves to protect your hands); mulch bare soil to prevent more seeds from sprouting. Dig older plants, getting as much of the root system as possible. Or cut plants to the ground and cut down or pull new shoots every week or two, which will weaken, and then kill, the plants. Or mulch with cardboard or a thick layer of newspaper for at least one growing season.

- **Common chickweed**

    DESCRIPTION: Common chickweed *(Stellaria media)* is an annual weed that spreads mainly by seed. It forms dense mats of many-branched, trailing or weakly upright stems that root where they touch the soil. Plants may be up to 12 inches tall but are often much lower. The stems bear pairs of small oval leaves with pointed tips. Tiny white flowers bloom near the ends of the stems from early spring through fall.

    CONTROL: Hoe or pull seedlings; mulch bare soil to prevent more seeds from sprouting. Pull, cut, or mow larger plants before they flower and set seed.

■ Crabgrasses

Description: Crabgrasses (*Digitaria* spp.) are grassy, annual weeds that spread by seed. Branching, thick, spreading stems carry coarse blue-green to purplish leaf blades that may be smooth or hairy. From summer into fall, plants are topped with narrow, fingerlike spikes of inconspicuous florets. Plants may grow up to 3 feet tall but adapt quickly to whatever mowing height you use.

Control: Pull or dig plants before they set seed. In lawns, discourage crabgrass from returning by taking steps to improve lawn vigor. Fertilize and lime as needed, and water deeply only when the grass shows signs of wilting. Avoid mowing too low; 2 inches is fine for most lawns. Thick, healthy turf will shade out crabgrass and discourage seeds from sprouting.

■ Curly dock

DESCRIPTION: Curly dock *(Rumex crispus)* is a perennial weed that reproduces by seed. It grows from a large, somewhat branched, yellowish taproot. Long, smooth, dark green leaves are lance-shaped with a rounded or heart-shaped base. The smooth, upright flower stems grow up 4 feet tall; they bear few leaves. From summer into fall, small flowers grow in dense clusters along the ends of the stems. They are greenish when young but turn reddish brown as they mature.

CONTROL: Dig out plants, getting as much of the taproot as possible, or cut them to the ground. Cut regrowth every week or two, or mulch with cardboard or a thick layer of newspaper for one growing season.

- **Dandelion**

    DESCRIPTION: Dandelion *(Taraxacum officinale)* is a perennial weed that spreads mainly by seed. Rosettes of serrated or straight-edged leaves grow from thick taproots with branching crowns. Each 2- to 12-inch hollow flower stem is topped with a 1- to 2-inch–wide bloom that is packed with narrow, strap-shaped, yellow petals. These flowers mature into puffy white seedheads. Plants may bloom any time from spring to frost, or year-round in warm areas.

    CONTROL: Dig plants as soon as you spot them, getting as much of the taproot as possible, or cut plants to the ground. Cut regrowth every week or two, or mulch with cardboard or a thick layer of newspaper for one growing season.

- **Foxtails**

    DESCRIPTION: Foxtails (*Setaria* spp.) are grassy, annual weeds that reproduce by seed. The upright or creeping stems branch at the base and bear flat green leaf blades. From summer into fall, the stems are topped with hairy greenish to yellowish spikes that may be upright or nodding. Plants may grow up to 3 feet tall but can grow much lower in mowed areas.

    CONTROL: Hoe or pull seedlings; mulch bare soil to prevent more seeds from sprouting. Dig out older plants or cut them to the ground every week or two until no new sprouts appear.

- **Ground ivy**

    DESCRIPTION: Ground ivy *(Glechoma hederacea)* is a perennial weed that reproduces by seed and by creeping stems that root as they spread. They may grow several feet each season. The whole plant is usually less than 3 inches tall. The square stems bear pairs of bright green, rounded leaves with scalloped edges. Small purplish flowers bloom along the stems in spring. Plants have a minty odor when crushed.

    CONTROL: Pull stems and roots as soon as you notice the plants; repeat every week or two to control regrowth. Regular fertilizing and aerating can improve lawn vigor and discourage the weed from returning. To reclaim heavily infested areas, pull or mow plants, then mulch with cardboard or a thick layer of newspaper for at least one growing season.

■ Henbit

DESCRIPTION: Henbit *(Lamium amplexicaule)* is an annual or biennial weed that reproduces mainly by seed. The square stems tend to be sprawling and root where they touch the soil; upright branches to 1 foot tall arise from the main stems. The hairy, rounded, scalloped leaves are borne in pairs; the bases of the upper leaves clasp the stem. Pinkish to purple flowers bloom in clusters near the tops of the upright stems in spring or fall.

CONTROL: Hoe or pull seedlings; mulch bare soil to discourage more seeds from germinating. Dig or pull larger plants before they set seed.

■ Lamb's-quarters

DESCRIPTION: Lamb's-quarters *(Chenopodium album)* is an annual weed that reproduces by seed. The upright, branching stems often have reddish streaks; they grow up to 3 feet tall. The smooth leaves are oval to lance-shaped or triangular; their edges may be slightly toothed. The blue-green foliage may be covered with a powdery white material, especially on the undersides. From summer into fall, inconspicuous greenish flowers bloom in elongated clusters along upper parts of the stems.

CONTROL: Hoe or pull seedlings; mulch to prevent more seeds from sprouting. Dig out older plants or cut them to the ground every week or two until no new sprouts appear.

■ Nimblewill

DESCRIPTION: Nimblewill *(Muhlenbergia schreiberi)* is a grassy perennial weed that reproduces mainly by seed. The slender, branching stems may also root at the lower leaf joints; they tend to grow more upright toward the tips, with flat and slender leaf blades. In fall, the stems are topped with thin spikes of inconspicuous green florets that turn reddish brown as they mature. Plants may reach 2 feet tall but can grow lower in mowed areas.

CONTROL: Dig out plants and their roots, or hoe or cut stems to the ground. Repeat every week or two to control regrowth, or mulch with cardboard or a thick layer of newspaper for one growing season.

- **Prostrate knotweed**

    DESCRIPTION: Prostrate knotweed *(Polygonum aviculare)* is an annual weed that reproduces by seed. The branched stems generally creep along the ground but may turn upward at the ends; plants are usually no more than 12 inches tall. The stems bear lance-shaped to oblong, pointed, blue-green leaves that turn reddish brown after a killing frost. Tiny white or yellowish flowers bloom along the stems from midsummer through fall.

    CONTROL: Hoe or pull seedlings; mulch bare soil to prevent more seeds from sprouting. Pull or cut older plants to the ground before they set seed, being sure to get the crown (where the stems join the roots). Prostrate knotweed thrives in compacted, high-traffic areas; prevent its return by adding organic matter to loosen the soil and by aerating lawn areas yearly.

■ Prostrate spurge

DESCRIPTION: Prostrate spurge *(Euphorbia maculata)* is an annual weed that reproduces by seed. The slender reddish or purplish stems branch freely; they may creep or turn upright at the tips. Plants form broad mats that are normally less than 6 inches tall. The greenish leaves are oblong to oval and usually have a purple-brown blotch on the top. Leaves, stems, and roots exude a milky sap when broken. Inconspicuous flowers bloom in clusters along stems from summer into fall.

CONTROL: Hoe, pull, or dig out plants before they set seed. Mulch bare soil to prevent more seeds from sprouting.

■ Quackgrass

DESCRIPTION: Quackgrass *(Agropyron repens)* is a grassy, perennial weed reproducing by seeds and fast-creeping roots (rhizomes). The narrow, blue-green leaves are soft and flat, their bases clasp the stem. Narrow spikes of green florets bloom atop the stems from summer into early fall. Plants may reach 4 feet tall but can grow lower in mowed areas.

CONTROL: Dig out the roots and slender, pointed rhizomes as thoroughly as possible; be on the lookout for new shoots from missed roots. Or hoe or cut plants to the ground every week or two until no new shoots appear. In severely infested areas, cut or mow plants, then mulch with cardboard or newspaper for at least one growing season.

- ## Shepherd's-purse

    DESCRIPTION: Shepherd's-purse *(Capsella bursa-pastoris)* is an annual weed that reproduces by seed. The leaves may be coarsely toothed or lobed; they grow mainly in a low rosette. Slender, upright, sparsely leaved stems grow to 30 inches tall. Small white four-petaled flowers bloom at the tips from early spring through late fall. They mature into small triangular seedpods.

    CONTROL: Hoe or pull seedlings; mulch bare soil to prevent more seeds from sprouting. Pull or cut older plants to the ground before they flower; repeat every week or two to control regrowth.

- ## Wild garlic

    DESCRIPTION: Wild garlic *(Allium vineale)* is a perennial weed that reproduces by seed and bulblets. Its underground bulbs produce stiff, upright, leafy stems. The slender leaves taper at the tip and wrap around the stem at the base. Stems may be topped with a rounded cluster of greenish to purplish, small flowers or tiny bulblets in late spring to early summer. Wild garlic can reach 3 feet tall but is lower in mowed areas. All parts of the plant have a strong onion or garlic odor when crushed.

    CONTROL: Dig clumps, being sure to get as many of the bulbs as possible. Or cut plants to the ground every week or two, until no new growth appears. In heavily infested areas, cut or mow plants, then mulch with cardboard or a thick layer of newspaper for at least one growing season.

- **Yellow nutsedge**

    DESCRIPTION: Yellow nutsedge *(Cyperus esculentus)* is a perennial weed that reproduces by seed and by thin creeping stems tipped with hard, small tubers. The upright, three-sided, 1-foot stems bear pale green grasslike leaves; the leaf bases wrap around the stems. Plants are topped with loosely branched, scaly, golden brown spikes from summer into fall.

    CONTROL: Pull or cut down plants as soon as you see them. Keep removing new shoots every week or two until no more sprouts appear; persistence is the key. In severely infested areas, cut or mow plants, then mulch with cardboard or a thick layer of newspaper for at least one growing season.

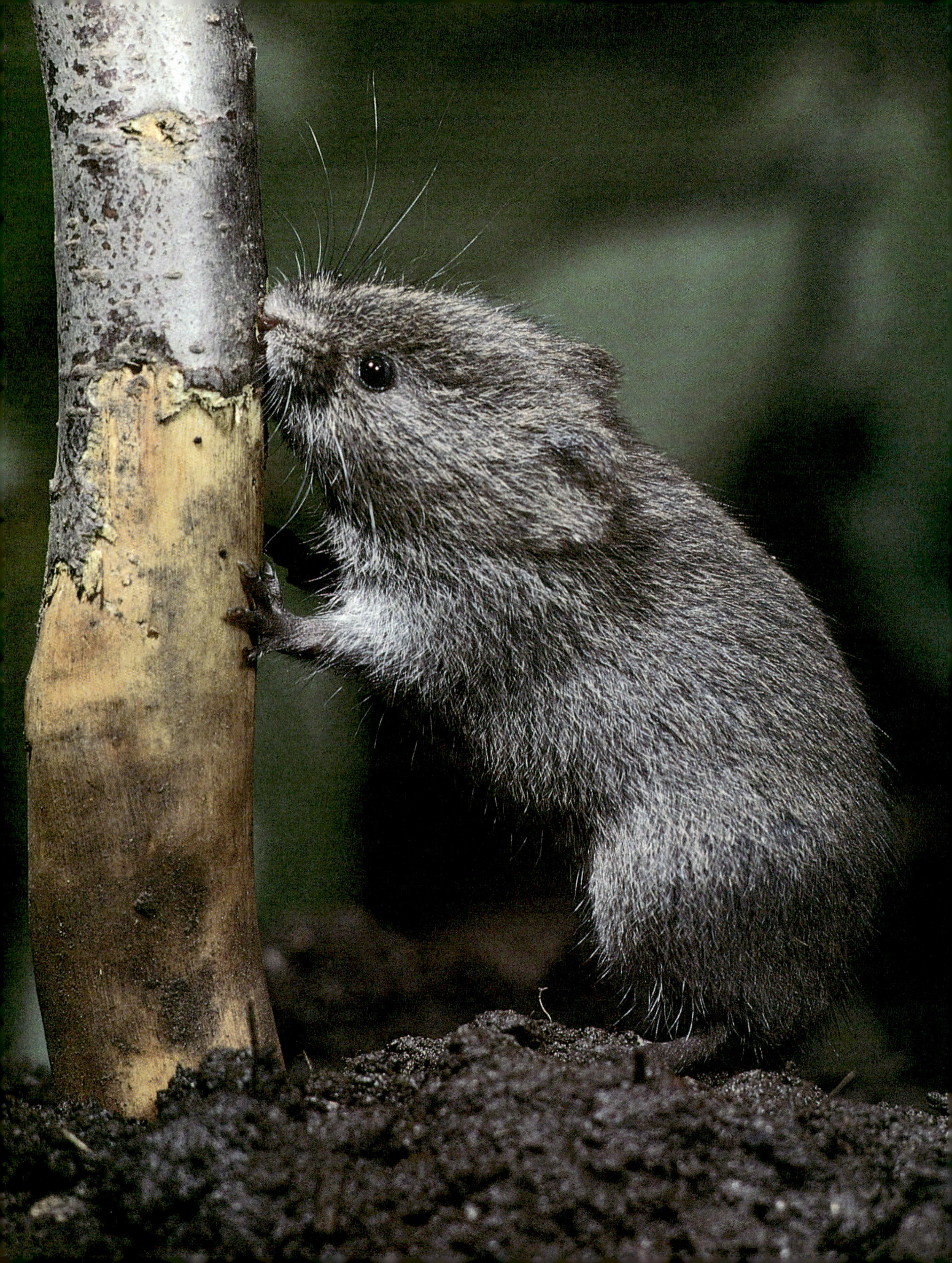

Chapter 5:
Handling Animal Pests

Few garden problems are as serious — or appear as suddenly — as the damage caused by animal pests. Within seconds, animals can crush whole plants. Their feeding can damage or demolish leaves, stems, and buds, ruining the look of a plant for that year or weakening it to the point of death.

Because the damage can happen so quickly, prevention is the key to protecting your plants from animal pests. In some cases, repellents may be enough to encourage animals to go elsewhere. For more security, temporary or permanent fences are the way to go. Read on to find details on your options for coping with deer, rabbits, and other common pests.

Dealing with Deer

Deer are a serious garden problem in many areas. A few hungry deer can quickly demolish even long-established garden plants. If you only occasionally see deer

Especially in winter, hungry rodents like this meadow vole will gnaw through the bark of trees or shrubs, girdling the stems. A collar of hardware cloth will keep them at bay.

around your property, repellents might be enough to discourage them from feeding. Some gardeners have had good results with bars of soap or mesh bags of human hair hung from trees or stakes. You can also experiment with sprays of eggs, garlic, or hot pepper mixed with water, or buy a commercial repellent.

When deer are hungry and food is scarce, even strong repellents may not be enough to keep them away. At this point, fences are the answer. You can build cages around individual plants or enclose your whole yard. Deer are excellent high jumpers, so you need to make the fence at least 8 feet tall. Or you could take advantage of their poor long-jumping ability and try two 4-foot fences spaced 3 feet apart.

Keeping Out Rabbits

Rabbits cause most of their damage feeding on tender young plants in spring. But they can also severely injure trees and shrubs in winter by chewing the bark near the base of the stems. Discourage this winter damage by surrounding the base of woody stems with a collar of 1/4-inch mesh hardware cloth. Make the collar at least 1 foot tall, or extend it up to the first side branch in deep-snow areas. Leave a 2-inch gap all around between the trunk and the wire. To prevent damage in spring, protect young plants with floating row covers, or enclose the whole garden with a wire-mesh fence that extends at least 3 inches below ground.

Managing Mice, Voles, and Moles

Mice and voles are tiny but destructive garden pests. They will eat many kinds of plants and are especially fond of bulbs and tubers. During the winter, they may feed on the bark of trees and shrubs, girdling the stems and killing the plants. Prevent damage to woody plants with hardware-cloth collars, as discussed in "Keeping Out Rabbits" above. (Be sure to press the base of the collar into the soil to prevent mice and voles from crawling underneath.) Floating row covers or hardware-cloth screens can protect seedlings until they get established. Avoid applying winter mulches until the ground has frozen to discourage these pests from making a home in your beds.

Moles are often unjustly blamed for the damage caused by mice and voles.

Although mice and voles use mole tunnels to travel around the garden, the moles themselves feed only on earthworms, grubs, and other insects in the soil. However, many gardeners are bothered by the ridges and holes mole tunnels can create in lawn and garden areas. Applying milky disease spores to the soil can kill grubs and encourage moles to go elsewhere. To discourage moles from tunneling into an area, surround it with a 2-foot deep and 6-inch wide trench filled with gravel.

BAFFLING BIRDS

Most times, birds are a welcome addition to the garden, providing sound and beauty while helping to control insect pests. They can, however, become pests when they feed on your ripening fruit and vegetable crops.

Some gardeners have luck with scare tactics, such as bobbing balloons, plastic grocery bags attached to stakes so the bags flap in the breeze, or pie pans that rattle in the wind. But the only sure-fire way to prevent damage is to physically keep birds from reaching the plants. Protect crops with floating row covers or plastic netting. Netting also works well on fruiting bushes and trees. It's best to surround these plants with tall stakes or wooden or pipe frames, then drape the netting over the frames; otherwise, birds may be able to reach through the netting to pick at the fruit. Also make sure the base of the netting is secured to the ground or to the base of the plant to keep birds from hopping underneath.

HANDLING OTHER ANIMAL PESTS

Depending on where you live, you may have other animals to contend with. If you are new in an area, ask your neighbors about the kinds of animals that are commonly seen in your area so that you'll know what preventive steps to take.

Woodchucks. Also commonly known as groundhogs, these plump brown creatures can be a real terror in country and suburban gardens. They feed mainly in early morning or late afternoon and can quickly demolish whole plants, especially in spring. Sometimes, covering young plants with floating row covers until they get established is enough to prevent damage. If woodchucks are a persistent

Animal-proof fence

This simple fence will deter a variety of animal pests. Start by digging a 6-inch-deep, 6-inch-wide trench around your garden. Line the trench with fine-mesh hardware cloth or chicken wire. (Run the trench and the hardware cloth across where the garden gate will be.) Use 4-foot metal or rot-resistant wood posts for the fence itself, and attach fencing, hardware cloth, or chicken wire to both the posts and the top of the mesh in the trench. Do not attach the top foot of fencing to the posts; leaving it loose will deter climbing pests. When you add a gate, be sure you leave no more than a 1-inch gap between the gate and the ground.

6 in.

6 in.

problem in your area, surround your garden with a fence. To prevent these pests from tunneling underneath, dig a 1-foot deep and wide trench along the outer side of the fence, and line the trench with chicken wire. Fasten the top of the buried wire to the bottom of the existing fence, then fill in the trench.

Raccoons. These pesky pests have a sweet tooth for sweet corn, leaving broken stalks and empty, shredded husks in their wake. Raccoons climb easily, so a simple fence won't keep them out. But you can make an existing fence more effective by topping it with a strand of electric fence. If you don't have a fence, covering each ear with a paper bag and closing the base of the bag around the stem with masking tape may be enough to prevent damage.

Cats and Dogs. Even your beloved pets can wreak havoc in the garden. Cats may dig up seedlings, roll on plants, or shred tender stems with their claws. Dogs often fail to distinguish between their in-bounds lawn areas and out-of-bounds beds, trampling whole plantings with one frolic through the garden. In either case, fencing off garden areas is a sensible way to prevent damage. Cylinders made from wire mesh and placed around the stems of young trees and shrubs can keep cats from sharpening their claws there. Shielding young plants with floating row covers or screens made from wire mesh can also prevent pets from digging up or crushing seedlings.

Hardiness Zone Map

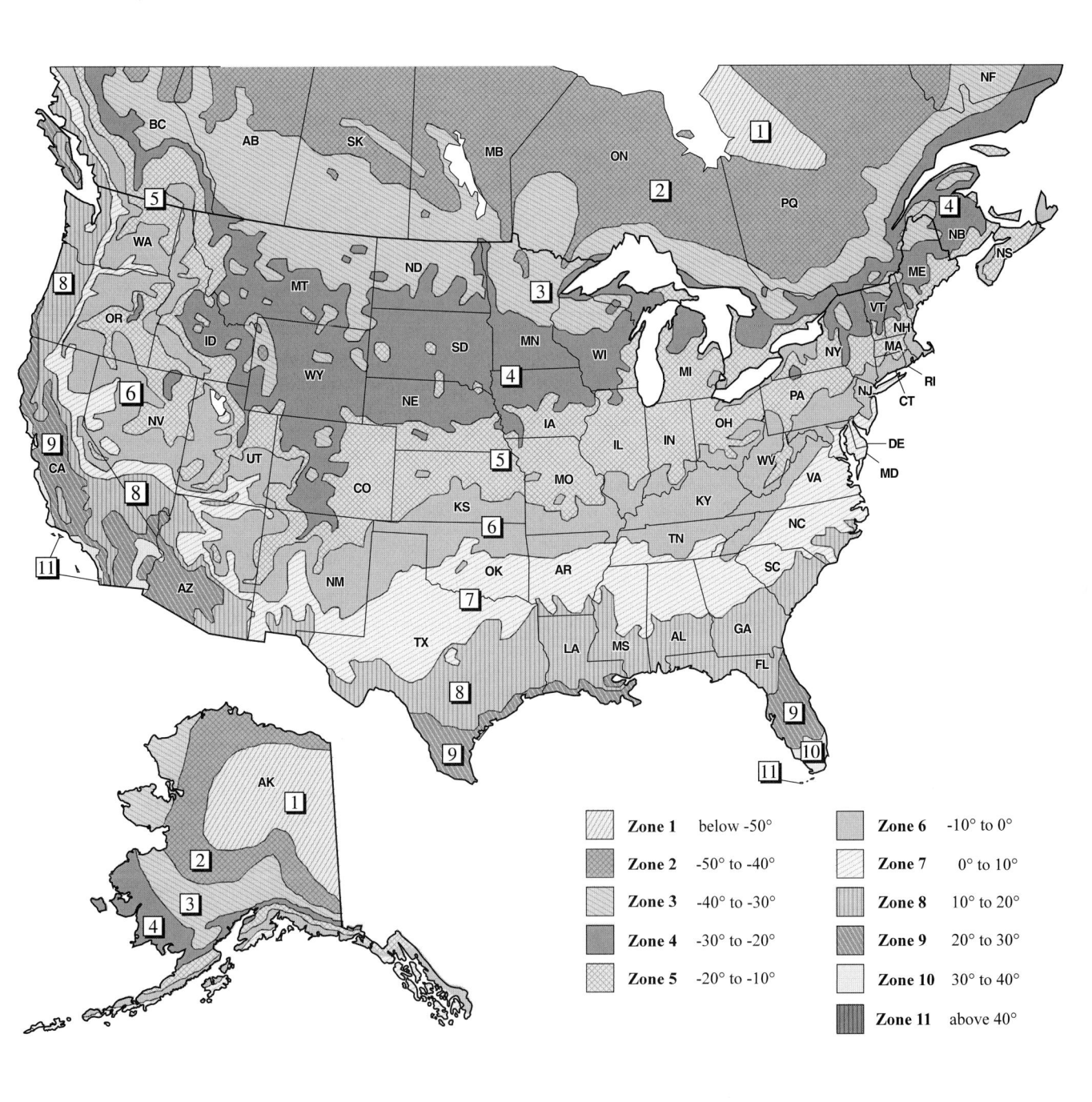

Zone	Temperature
Zone 1	below -50°
Zone 2	-50° to -40°
Zone 3	-40° to -30°
Zone 4	-30° to -20°
Zone 5	-20° to -10°
Zone 6	-10° to 0°
Zone 7	0° to 10°
Zone 8	10° to 20°
Zone 9	20° to 30°
Zone 10	30° to 40°
Zone 11	above 40°

Sources

Soil Testing Laboratories

A & L Agricultural Labs
7621 White Pine Road
Richmond, VA 23237

Biosystem Consultants
P.O. Box 43
Lorane, OR 97541

Timberleaf
5569 State Street
Albany, OH 45710

Wallace Labs
365 Coral Circle
El Segundo, CA 90245

Pest and Disease Control Equipment and Supplies

The following companies offer a good selection of garden tools and organic controls including beneficial insects, traps, sprays, and dusts.

Bountiful Gardens
19550 Walker Road
Willits, CA 95490

Gardener's Supply Co.
128 Intervale Road
Burlington, VT 05401

Harmony Farm Supply
P.O. Box 460
Graton, CA 95444

Peaceful Valley Farm Supply Co.
P.O. Box 2209
Grass Valley, CA 95945

Photo Credits

Cathy Wilkinson Barash: 33

David Cavagnaro: 2, 87

E. R. Degginger: back cover

Wally Eberhart/Photo/Nats: 35, 76

Barbara W. Ellis: 13, 29

Derek Fell: vi–1, 6, 23, 89

Charles Marden Fitch: 30

Dwight R. Kuhn: iii, 11, 12, 28, 31, 108, back cover

Jerry Pavia: 68

Michael S. Thompson: 84

INDEX

Achillea, 12
Ageratum, 14
Agropyron repens, 104
Alder, leafminers and, 52
Allium vineale, 106
Animal pests, 109–13
Annual flowers and vegetables, 7
Aphids, 12, 16, 22, 30, 36
Apple, 9
   brown rot and, 72
   canker and, 72
   cedar-apple rust and, 73
   crown gall and, 74
   fire blight and, 77
   leafrollers and, 53
   plum curculios and, 60
   tent caterpillars and, 64–65
Apple maggots, 36–37
Apple scab, 70
Appletree borers, 40
Apricot, brown rot and, 72
Armyworm, 23, 37
Asparagus, rust and, 82
Asparagus beetles, 39
Aster, plant bugs and, 59
Azalea, lace bugs and, 51

Bacillus thuringiensis. See BT
Bacillus thuringiensis var. *kurstaki.* See BTK
Bacillus thuringiensis var. *san diego.* See BTSD
Bacteria, 22, 28
Bacterial disease, 33. *See also* Bacteria; Diseases
Bagworm, 38
Bean, 14
   rust and, 82
   stink bugs and, 64
   thrips and, 65

Bee balm, mildew and, 80
Beet, 14
   leafminers and, 52
Beet family, 14
Beetle, 22, 25, 39
   asparagus, 39
   Colorado potato, 43
   cucumber, 45
   flea, 46–47
   Japanese, 39
   lady beetles, 30, 31
   larvae of, 35
   Mexican bean, 54–55
   rose chafers, 39
Beneficial insects, 10–13, 25–28. *See also* insects by name
Bentgrass, sod webworms and, 62
Bindweed, 90
Biological controls, 25–29
Birch, leafminers and, 52
Birds, 10, 111
Black cherry, tent caterpillars and, 64–65
Black spot, 70
Black vine weevil, 66
Blight, 22
   fire, 77
Blossom end rot, 71
Blueberry, plum curculios and, 60
Bluegrass, sod webworms and, 62
Bordeaux mix, 33
Borer, 22, 40
Botanical family, crop rotation and, 14
Botanical pesticides, 32
Bramble
   crown gall and, 74
   rust and, 82
Broadleaf plantain, 91

Broad-spectrum pesticides, 32
Brown rot, 72
Brussels sprouts, 14
BT, 1, 4, 29
BTK, 37, 41–42, 48
BTSD, 43
Budworm, spruce, 63
Bugs. *See* Insects; Lace bug; Mealybug; Plant bug; Stink bug

Cabbage, 14
   Harlequin bugs and, 64
   leafminers and, 52
   weevils and, 66
Cabbage family, 14
   club root and, 74
Cabbage looper, 41
Cabbageworm, 23, 41
Calendula, 14
Canada thistle, 92
Canker, 72
Cankerworm, 41
Capsella bursa-pastoris, 105
Caraway, 12
Cardboard collar, 26
Carrot, 14
   weevils and, 66
Carrot family, 12, 14
Caterpillar, 27, 41–42
   codling moths, 42–43
   leaf-feeding, 41
   tent, 64–65
Catnip, 12
Cats, 113
Cauliflower, 14
Cedar-apple rust, 73
Celery, 14
Chenopodium album, 100
Cherry
   brown rot and, 72
   plum curculios and, 60

Cherry fruit flies, 47
Chewing insects, 22
Chickweed, common, 93
Chives, 14
Chrysanthemum, leafminers and, 52
Chrysanthemum × morifolium, 12
Cirsium arvense, 92
Clandosan, 21
Climate, 9
Club root, 74
Codling moth, 27, 42–43
Collard, 14
Collars
   cardboard, 26
   cutworm, 26
   hardware cloth, 109
Colorado potato beetles, 43
Columbine, leafminers and, 52
Common chickweed, 93
Compost pile, 20
Coneflower, 12
Conifer, spruce budworms and, 63
Conifer sawflies, 44
Controls, 24–33
Convolvulus spp., 90
Cooperative extension service, 24
Copper, 33
Copper sulfate, 33
Corn, 14
   raccoons and, 113
Corn borer, 23, 40
Corn earworm, 41
Cosmos, 14
Crabapple, 9–10
   apple scab and, 70
   cedar-apple rust and, 73
Crabgrass, 94
Cricket, mole, 56

Index

Crop
  grubs and, 49
  mealybugs and, 54
  for weed control, 86
  See also Animal pests
Crop rotation, 14–15
Crown gall, 74
Cucumber, 14
  mildew and, 80
Cucumber beetle, 45
Cultural controls, 25
Curculios, plum, 60
Curly dock, 95
Currant fruit flies, 47
Cutworm, 22, 45
Cutworm collar, 26
Cyperus esculentus, 107

Dahlia, plant bugs and, 59
Daisy, 12
  Shasta, 12
Daisy family, 12, 14
Damaged plants, 20–24
Damping-off, 75
Dandelion, 96
Deciduous trees
  bagworms and, 38
  gypsy moths and, 50
  tent caterpillars and, 64–65
Deer, 109–10
Deficiency, nitrogen, 80
Diatomaceous earth, 31–32
Digitaria spp., 94
Dill, 12, 14
Diseases, 69–83
  discouraging, 14–15
  sprays, dusts, and, 30
  See also diseases by name
Dogs, 113
Downy mildew.
  See Mildew
Dusts, 1, 30–33

Eastern red cedar, cedar-apple rust and, 73
Echinacea, 12

Edging strips, 89
Edible plants
  gray mold and, 78
  leaf spots and, 79
  mildew and, 80
  rust and, 82
  viruses and, 82
  wilts and, 83
Eggplant, 14
  Colorado potato beetles and, 43
Elm, leafminers and, 52
Equipment, for pest and disease control, 117
Euonymus, crown gall and, 74
Euphorbia maculata, 103
Evergreens
  bagworms and, 38
  gypsy moths and, 50
  See also evergreens by name

Fairy ring, 76
Fall webworm, 46
Fence, animal proof, 112–113
Fennel, 12, 14
Fertilizing, 18
Fir, spruce budworms and, 63
Fire blight, 77
Firethorn, fire blight and, 77
Flaming, 88
Flea beetle, 46–47
Floating row covers, 28, 111, 113
Flowering tobacco, 14
Flowers
  aphids and, 36
  beetles and, 39
  cutworms and, 45
  lace bugs and, 51
Fly, fruit, 47.
  See also Sawfly
Forsythia, crown gall and, 74

Foxtail, 97
Frogs, 10
Fruit
  aphids and, 36
  codling moths, 42–43
  leafhoppers and, 51
  mealybugs and, 54
  mites and, 55
  plum curculios and, 60
  See also fruit by type
Fruit borer, 40
Fruit fly, 47
Fruit trees
  brown rot and, 72
  oriental fruit moths and, 56
  pear sawflies and, 58
Fungal disease, 33
Fungi, 22, 28.
  See also Diseases; fungi by name
Fungicidal soaps, 30

Gall, crown, 74
Garden crops, grubs and, 49. *See also* Crop
Garden webworm, 48
Garlic, 14
  wild, 106
Gladiolus
  thrips and, 65
  wireworms and, 67
Glechoma hederacea, 98
Goldenrod, 11
Grape, mildew and, 80
Grass, 89
  armyworms and, 37
  crabgrass, 94
  grubs and, 49
  quackgrass, 104
  sod webworms and, 62
  See also Lawn grass; Turf grass
Grass family, 14
Gray mold, 78
Green lacewing, 25
Grooming, 18–19
Ground ivy, 98

Grubs, 49
Gypsy moth, 50
  larvae of, 27

Hardiness zone map, 115
Harlequin bug, 64
Hemlock, spruce budworms and, 63
Henbit, 99
Herbicides, 88.
  See also Insecticides
Herbs, aphids and, 36
Hoeing, 86
Honeydew, 22
  aphids and, 36
  leafhoppers and, 51
  mealybugs and, 54
  whiteflies and, 67
Hornworm, tomato, 23
Horticultural oils, 1, 31
Houseplants
  mealybugs and, 54
  mites and, 55

Information sources, 117
Insecticidal soaps, 30
Insecticides,
  microbial, 28–29.
  See also Herbicides
Insects, 16, 30
  scale, 60–61
  See also Beneficial insects; Pests; insects by name
Iris borer, 40
Irrigation. *See* Watering
Ivy, ground, 98

Johnny jump-up, 87
Juniper, cedar-apple rust and, 73

Kale, 14
Knotweed, prostrate, 102
Kohlrabi, 14

Lace bug, 51
Lacewing, 12, 25

Index

Lady beetle, 25, 30
  larvae of, 31
Lamb's-quarters, 100
Lamium amplexicaule, 99
Larvae, as biological controls, 25
Lawn grass, 89
  armyworms and, 37
  grubs and, 49
  mole crickets and, 56
Leaf curl, peach, 81
Leaf scorch, 78
Leaf spots, 22, 30, 79
Leafhopper, 51
Leafminer, 22, 52
Leafroller, 53
Lettuce, 14
  mildew and, 80
Lilac, mildew and, 80
Lime, 33

Maggots, 27
  apple, 36–37
Mantid. *See* Praying mantid
Maple, canker and, 72
Marigold, 14
Meadow vole, 108
Mealybug, 16, 22, 30, 54
Mechanical controls, 25, 26–27
Mediterranean fruit fly, 47
Melon, 14
  cucumber beetles and, 45
  mildew and, 80
Mexican bean beetle, 54–55
Mice, 110–11
Microbial insecticides, 28–29
Microclimates, 9
Mildew, 22, 30, 80
Minerals, 33
Mint, 12
  plant bugs and, 59
Mint family, 12
Mites, 22, 55

Mold, 22
  gray, 78
Mole cricket, 56
Moles, 110–11
Moths, 27
  codling, 42–43
  gypsy, 50
  oriental fruit, 56–57
Muhlenbergia schreiberi, 101
Mulch, 3, 16–17, 86
  organic, 16

Nasturtium, leafminers and, 52
Nectarine, peach leaf curl and, 81
Neem, 32
Nematode, 21
  parasitic, 35
Netting, 111
Nimblewill, 101
Nitrogen deficiency, 80
Nutrients, soil, 7
Nutsedge, yellow, 107

Oils. *See* Horticultural oils
Okra, stink bugs and, 64
Onion, 14
  thrips and, 65
Onion family, 14
Oregano, 12
Organic dusts and sprays. *See* Dusts; Sprays
Organic matter, 7–8
Organic mulch, 16
Oriental fruit moth, 56–57
Ornamentals
  gray mold and, 78
  leaf spots and, 79
  leafhoppers and, 51
  mealybugs and, 54
  mildew and, 80
  mites and, 55
  oriental fruit moths and, 56
  rust and, 82

viruses and, 82
weevils and, 66
whiteflies and, 67
wilts and, 83

Parasitic nematodes, 35
Parasitoids, 25
Parsley, 12, 14
Parsnip, 14
Pathogens, 25
  See also Microbial insecticides
Pea, 14
  stink bugs and, 64
Pea family, 14
Peach
  brown rot and, 72
  canker and, 72
  plant bugs and, 59
  plum curculios and, 60
  stink bugs and, 64
Peach leaf curl, 81
Peachtree borer, 40
Pear
  brown rot and, 72
  canker and, 72
  crown gall and, 74
  fire blight and, 77
  plum curculios and, 60
  tent caterpillars and, 64–65
Pear sawfly, 58
Pepper, 14
  blossom end rot and, 71
Perennials, 7
Pesticides
  botanical, 32
  microbial, 29–32
  See also Herbicides
Pests, 35–67
Petunia, 14
  Colorado potato beetles and, 43
pH, 7, 8
Pheromones. *See* Traps
Phlox, mildew and, 80
Physical controls, of weeds, 88. *See also* Mechanical controls

Phytotoxic sprays and dusts, 30
Pine, spruce budworms and, 63
Plant bug, 59
Plantago major, 91
Plantain, broadleaf, 91
Plantings, diverse, 10–14
Plum
  brown rot and, 72
  canker and, 72
  tent caterpillars and, 64–65
Plum curculio, 60
Polygonum aviculare, 102
Potato, 14
Powdery mildew. *See* Mildew
Praying mantid, 29
Predators, 25
Prostrate knotweed, 102
Prostrate spurge, 103
Pruning, 18–19
Pupae, 23
Pyrethrins, 32

Quackgrass, 104
Queen-Anne's lace, 1

Rabbits, 110
Raccoons, 113
Radish, 14
  weevils and, 66
Records, 20
Resistance, 9–10
Rhododendron, black vine weevils and, 66
Rodents, 108
Roots, beetles and, 39. *See also* Club root
Rose, 10, 14
  black spot and, 70
  canker and, 72
  crown gall and, 74
  leafrollers and, 53
  mildew and, 80
  rust and, 82
Rose chafers, 39

Index

Rose family
  fire blight and, 77
  pear sawflies and, 58
Rosemary, 12
Rot, 22, 30, 81
  blossom end, 71
  brown, 72
Rotation system, 14–15
Rotenone, 1
Roundworms. *See* Nematode
Row covers, 28, 111, 113
Rudbeckia spp., 12
Rumex crispus, 95
Rust, 30
  cedar-apple, 73

Sabadilla, 1, 32
Sage, 12
Sanitation, 19–20
Sawfly
  conifer, 44
  pear, 58
Scab, apple, 70
Scale, 22
Scale insects, 16, 30, 60–61
Scorch, leaf, 78
Screens, 113
Seed, damping-off fungi and, 75
Seedlings, 15
  damping-off fungi and, 75
  See also Animal pests
Setaria spp., 97
Shallot, 14
Shasta daisy, 12
Shepherd's-purse, 105
Shrubs
  canker and, 72
  lace bugs and, 51
  tent caterpillars and, 64–65
Slugs, 17, 22, 25, 61–62. *See also* Sawfly, pear
Snails, 17, 22, 25, 61–62
Snapdragon, rust and, 82
Soaps, 1. *See also* Sprays
Sod webworms, 62

Soft controls, 1, 4
Soil, 4–8
  nutrient levels, 7
  pH level, 7
  tests of, 5–7
Soil testing laboratories, 117
Solarizing, 86
Sources of information, 117
Sphagnum moss, 15
Spider, golden garden, 13
Spider mite, 30
Spinach
  leafminers and, 52
  mildew and, 80
Spirea, leafrollers and, 53
Spots. *See* Black spot
Sprays, 1, 30–33
  safe use of, 32
  for weed control, 88
Spruce budworm, 63
Spurge, prostrate, 103
Squash, 14
  cucumber beetles and, 45
  mildew and, 80
  thrips and, 65
Squash borers, 40
Squash family, 14
Stalk borer, 40
Stellaria media, 93
Stems, beetles and, 39
Stink bug, 64
Strawberry
  garden webworms and, 48
  leafrollers and, 53
  plant bugs and, 59
Strawberry weevil, 66
Sucking insects, 22, 30
Sulfur, 33
Supplies, for pest and disease control, 117
Sweet alyssum, 14
Sweet corn, raccoons and, 113
Sweet pea, 14
Swiss chard, 14

leafminers and, 52
mildew and, 80
plant bugs and, 59
weevils and, 66

Tanglefoot, 27, 37
Taraxacum officinale, 96
Tent caterpillars, 64–65
Thistle, Canada, 92
Thrips, 30, 65
Thyme, 12
Tilling, 86
Toads, 10
Tomato, 14
  blossom end rot and, 71
  stink bugs and, 64
Tomato family, 14
  Colorado potato beetles and, 43
Tomato hornworm, 23, 41
Transplanting, 17
Traps
  apple maggot, 27
  pheromone, 27
Tree band, 27
Trees
  aphids and, 36
  bagworms and, 38
  canker and, 72
  conifer sawflies and, 44
  fall webworms and, 46
  gypsy moths and, 50
  leafminers and, 52
  tent caterpillars and, 64–65
  See also Deciduous trees; Evergreens; Fruit trees
Turf grass
  fairy rings and, 76–77
  sod webworms and, 62
Turnip, 14
  Harlequin bugs and, 64

Vegetable weevils, 66
Vegetables
  aphids and, 36
  armyworms and, 37
  Colorado potato beetles and, 43

cutworms and, 45
flea beetles and, 46–47
garden webworms and, 48
lace bugs and, 51
leafhoppers and, 51
mealybugs and, 54
mites and, 55
nitrogen deficiency and, 80
stink bugs and, 64
whiteflies and, 67
wireworms and, 67
Viruses, 22, 24, 28, 82
Voles, 110–11

Walnut husk fruit flies, 47
Wasp, braconid, 23
Water garden, 10
Watering, 18
Webworm
  fall, 46
  garden, 48
  sod, 62
Weeds, 11, 85–107
  controlling, 87–88
  types of, 90–107
Weevil, 66
Whitefly, 16, 30, 66–67
Wild garlic, 106
Willows, canker and, 72
Wilt, 83
Wireworms, 67
Woodchucks, 111–13
Worms
  armyworms, 37
  bagworms, 38
  cutworms, 45
  spruce budworms, 63
  webworms, 46, 48, 62
  wireworms, 67

Yarrow, 11, 12
Yellow nutsedge, 107
Yew, black vine weevils and, 66

Zone map, 115

Titles available in the Taylor's Weekend Gardening Guides series:

Organic Pest and Disease Control	$12.95
Safe and Easy Lawn Care	12.95
Window Boxes	12.95
Attracting Birds and Butterflies	12.95
Water Gardens	12.95
Pruning	12.95

At your bookstore or by calling 1-800-225-3362

Prices subject to change without notice